Microscale Experiments

HOLT, RINEHART AND WINSTON
Harcourt Brace & Company

Austin • New York • Orlando • Atlanta • San Francisco • Boston • Dallas • Toronto • London

Copyright © by Holt, Rinehart and Winston
All rights reserved. No part of this publication may be reproduced or transmitted in any form or by any means, electronic or mechanical, including photocopy, recording, or any information storage and retrieval system, without permission in writing from the publisher.

Requests for permission to make copies of any part of the work should be mailed to the following address: Permissions Department, Holt, Rinehart and Winston, 1120 South Capital of Texas Highway, Austin Texas 78746-6487

Cover: HRW Photo by Sam Dudgeon

Printed in the United States of America

ISBN 0-03-051927-6

8 9 10 11 12 13 14 15   022   05 04 03 02 01

LAB MANUAL B CONTENTS

	Teacher's Safety Information	T5
	Master Materials List	T8
	Chemicals	T8
	Equipment	T11
	Miscellaneous Materials	T13
	Safety Equipment	T14
	Chemical Inventory	T15
	Incompatible Chemicals	T15
	Introduction	iv
	Safety	vi
	Labeling of Chemicals	xi
	Laboratory Techniques	xii
B1	Laboratory Procedures	1
B2	Accuracy and Precision: Calibrating a Pipet	21
B3	Relative Solubility of Transition Elements	29
B4	Periodicity of Properties of Oxides	33
B5	Reactivity of Halide Ions	39
B6	Chemical Bonds	43
B7	Conductivity as an Indicator of Bond Type	47
B8	Test for Iron(II) and Iron(III)	53
B9	Colored Precipitates	57
B10	Simple Qualitative Analysis	63
B11	Generating and Collecting O_2	67
B12	Generating and Collecting H_2	73
B13	Testing Reaction Combinations of H_2 and O_2	79

CONTENTS continued

B14 Reacting Ionic Species in Aqueous Solution **87**

B15 Hydronium Ion Concentration, pH **91**

B16 Titration of an Acid with a Base **95**

B17 Percentage of Acetic Acid in Vinegar **99**

B18 Clock Reactions **105**

B19 Equilibrium **109**

B20 Oxidation-Reduction Reactions **115**

B21 Cathodic Protection: Factors That Affect the Corrosion of Iron **119**

B22 Determination of Vitamin C in Fruit Juices **125**

Laboratory Safety

HELPING STUDENTS RECOGNIZE THE IMPORTANCE OF SAFETY

One method that can help students appreciate the importance of precautions is to use a "safety contract" that students read and sign, indicating they have read, understand, and will respect the necessary safety procedures, as well as any other written or verbal instructions that will be given in class. You can find a copy of a safety contract on the ***Holt ChemFile Teaching Resources CD-ROM.*** You can use this form as a model or make your own safety contract for your students with language specific to your lab situation. When making your own contract you could include points such as the following.

- Make sure that students agree to always wear personal protective equipment (goggles and lab aprons). Safety information regarding the use of contact lenses is continually changing. Check your state and local regulations on this subject. Students should agree to read all lab exercises before they come to class. They should agree to follow all directions and safety precautions, and use only materials and equipment that you provide.

- Students should agree to remain alert and cautious at all times in the lab. They should never leave experiments unattended.

- Students should not wear heavy dangling jewelry or bulky clothing.

- Students should bring lab manuals and lab notebooks only into the lab. Backpacks, textbooks for other subjects, and other items elsewhere should be stored elsewhere.

- Students should agree to never eat, drink, or smoke in any science laboratory. Food should NEVER be brought into the laboratory

- Students should NEVER taste or touch chemicals.

- Students should keep themselves and other objects away from Bunsen-burner flames. Students should be responsible for checking that gas valves and hot plates are turned off before leaving the lab.

- Students should know the proper fire drill procedures and the locations of fire exits.

- Students should always clean all apparatus and work areas.

- Students should wash their hands thoroughly with soap and water before leaving the lab room.

- Students should know the location and operation of all safety equipment in the laboratory.

- Students should report all accidents or close calls to you immediately, no matter how minor.

- Students should NEVER work alone in the laboratory and they should never work unless you are present.

SAFETY continued

DISPOSAL OF CHEMICALS

Only a relatively small percentage of waste chemicals are classified as hazardous by EPA regulations. The EPA regulations are derived from two acts (as amended) passed by the Congress of the United States: RCRA (Resource Conservation and Recovery Act) and CERCLA (Comprehensive Environmental Response, Compensation, and Liability Act).

In addition, some states have enacted legislation governing the disposal of hazardous wastes that differs to some extent from the federal legislation. The disposal procedures described in this book have been designed to comply with the federal legislation as described in the EPA regulations.

In most cases the disposal procedures indicated in the teacher's edition will *probably* comply with your state's disposal requirements. However, to be sure of this, check with your state's environmental agency. If a particular disposal procedure does not comply with your state requirements, ask that office to assist you in devising a procedure that is in compliance.

The following general practices are recommended in addition to the specific instructions given in the margin notes of this Teacher's Edition.

- Except when otherwise specified in the disposal procedures, neutralize acidic and basic wastes with 1.0 M potassium hydroxide, KOH, or 1.0 M sulfuric acid, H_2SO_4, added slowly while stirring.

- In dealing with a waste-disposal contractor, prepare a complete list of the chemicals you want to dispose of. Classify each chemical on your disposal list as hazardous or nonhazardous waste. Check with your local environmental agency office for the details of such classification.

- Unlabeled bottles are a special problem. They must be identified to the extent that they can be classified as a hazardous or nonhazardous waste. Some landfills will analyze a mystery bottle for a fee if it is shipped to the landfill in a separate package, is labeled as a sample, and includes instructions to analyze the contents sufficiently to allow proper disposal.

ELECTRICAL SAFETY

Although none of the labs in this manual require electrical equipment, several include options for the use of microcomputer-based laboratory equipment, pH meters, or other equipment. The following safety precautions to avoid electric shocks must be observed any time electrical equipment is present in the lab.

- Each electrical socket in the laboratory must be a three-hole socket and must be protected with a GFI (ground-fault interrupter) circuit.

- Check the polarity of all circuits before use with a polarity tester from an electronics supply store. Repair any incorrectly wired sockets.

SAFETY continued

- Use only electrical equipment equipped with a three-wire cord and three-prong plug.

- Be sure all electrical equipment is turned off before it is plugged into a socket. Turn off electrical equipment before it is unplugged.

- Wiring hookups should be made or altered only when apparatus is disconnected from the power source and the power switch is turned off.

- Do not let electrical cords dangle from work stations; dangling cords are a shock hazard and students can trip over them.

- Do not use electrical equipment with frayed or twisted cords.

- The area under and around electrical equipment should be dry; cords should not lie in puddles of spilled liquid.

- Hands should be dry when using electrical equipment.

- Do not use electrical equipment powered by 110-115 V alternating current for conductivity demonstrations or for any other use in which bare wires are exposed, even if the current is connected to a lower voltage AC or DC connection.

Use dry cells or Ni-Cad rechargeable batteries as direct current sources. Do not use automobile storage batteries or AC-to-DC converters; these two sources of DC current can present serious shock hazards.

Prepared by Jay A. Young, Consultant, Chemical Health and Safety, Silver Spring, Maryland

Master Materials List

Chemicals

Chemical	Amount needed for 15 lab groups	Experiment
Acetic acid solution, 5% (white vinegar)	80 mL	B17, B19
Acetic acid, glacial	6 mL	B15
Agar agar, powdered	30 g	B21
$AgNO_3$	0.85 g	B5
$Ba(NO_3)_2$	0.7 g	B14
$BaCl_2 \cdot 5H_2O$	15.6 g	B10, B14
$Ca(NO_3)_2$	4.1 g	B5
$CaCl_2$	30 g	B6
Chalk dust, yellow	60 g	B7
Citric acid	30 g	B6
$COCl_2 \cdot 6H_2O$	0.6 g	B9
$Cu(NO_3)_2 \cdot 3H_2O$	4.8 g	B3, B20
$CuCl_2 \cdot 2H_2O$	0.43 g	B9
$CuSO_4 \cdot 5H_2O$	6.25 g	B9, B12, B13, B19
Denatured alcohol	500 mL	B6, B16, B21, B22
$Fe(NH_4)_2(SO_4)_2 \cdot 6H_2O$	4 g	B8
$Fe(NO_3)_2$	15 g	B10
$Fe(NO_3)3 \cdot 6H_2O$	11 g	B3, B19
$Fe_3(SO_4)_2 \cdot 9H_2O$	1.4 g	B9
$FeCl_2$	15 g	B10
$FeCl_3 \cdot 6H_2O$	20 g	B8, B9, B10, B19, B20
$FeSO_4$	2 g	B20
Graphite, powdered	60 g	B7
H_2SO_4, concentrated	4 mL	B20, B21
H_3PO_4, concentrated	2 mL	B15
HCl, concentrated	152 mL	B4, B6, B9, B10, B12, B13, B15, B16, B17, B18, B19, B21
$Hg_2(NO_3)_2 \cdot 2H_2O$	5.6 g	B3
HNO_3, concentrated	12.8 mL	B3, B21
Hydrogen peroxide, 3%	1.5 L	B11, B13
Iodine (powdered)	0.6 g	B22

MASTER MATERIALS LIST continued

Chemical	Amount needed for 15 lab groups	Experiment
K_2HPO_4	7.5 g	B19
K_2SO_3	15 g	B10
$K_3Fe(CN)_6$	23.8 g	B8, B10, B21
$K_4Fe(CN)_6 \cdot 3H_2O$	29.4 g	B3, B8, B10
KBr	1.2 g	B5
KCl	22.5 g	B10, B19
KI	43.6 g	B5, B6, B11, B13, B22
$KMnO_4$	9.6 g	B10, B20
KNO_3	15.5 g	B10, B21
KOH	28 g	B15
KSCN	4.5 g	B3, B8, B19
Laundry starch	15 mL	B22
$LiNO_3$	15 g	B10
Methanol	100 mL	B19
Methyl red	0.5 g	B19
$Mg(NO_3)_2 \cdot 2H_2O$	1.8 g	B20
$MgCO_3$	15 g	B10
MgO	3.75 g	B4
$Na_2C_2O_4$	1.5 g	B14, B21
Na_2CO_3	32.5 g	B10, B15, B21
$Na_2S_2O_3$	255 g	B5, B10, B20, B22
Na_2SO_3	20 g	B10
Na_2SO_4	62.8 g	B3, B7, B14
Na_3PO_4	1.9 g	B21
$Na_3PO_4 \cdot H_2O$	38 g	B21
NaCl	63 g	B3, B5, B7, B14, B15, B21
NaF	0.5 g	B5
$NaHCO_3$	0.8 g	B15
$NaNO_3$	15.45 g	B10, B14
NaOCl solution, 5% (commercial bleach)	15 mL	B5
NaOH pellets	82.4 g	B3, B4, B8, B9, B10, B12, B13, B15, B16, B17, B19, B21
$NH_3(aq)$, (ammonia, water), concentrated	143 mL	B5, B15, B19
$(NH_4)_2C_2O_4 \cdot H_2O$	2.8 g	B3

ChemFile Lab B

MASTER MATERIALS LIST continued

Chemical	Amount needed for 15 lab groups	Experiment
$(NH_4)_2CO_3$	15 g	B10
NH_4CH_3COO	0.8 g	B15
NH_4Cl	20 g	B10
NH_4SCN	7.5 g	B19
$NiCl_2 \cdot 6H_2O$	0.6 g	B9
$NiSO_4 \cdot 6H_2O$	1.32 g	B9
Phenolphthalein	3 g	B16, B17, B21
Phenyl salicylate	30 g	B6
$SnCl_2$	2 g	B20
SnO_2	63.75 g	B4, B7
Sodium acetate (CH_3COONa)	15 g	B19
$SrCO_3$	7.5 g	B4
$SrNO_3$	15 g	B10
Starch (water-soluble)	4 g	B5, B22
Sucrose	30 g	B6
Sulfur	75 g	B4, B7
Tin	60 g	B7
Universal indicator	30 mL	B4
Vitamin C	50 mg	B22
Zinc	4.5 g	B12, B13
$Zn(NO_3)_2 \cdot 3H_2O$	48.6 g	B3, B10, B20
ZnO	3.75 g	B4
$ZnSO_4$	15 g	B10

MASTER MATERIALS LIST continued

Equipment

Equipment	Amount needed for 15 lab groups	Experiment
24-well microplate and lid	15	B3, B4, B5, B6, B16, B17, B19, B20, B21
Balance	15	B1, B2, B9, B18, B21
Beaker tongs	15	B18
Beaker, 100 mL	15	B2, B11, B12
Beaker, 250 mL	30	B1, B11, B12
Beaker, 400 mL	15	B21
Beaker, 50 mL	15	B1, B4, B18
Bunsen burner with gas tubing and striker (a hot plate may be substituted)	15	B1, B4, B10, B11, B12, B21
Buret clamp, double	15	B1
Burets	15	B1
Cobalt glass plate	30	B10
Conductivity tester, LED	15	B6, B7
Conductivity tester, LED	15	B7
Containers, 25 mL	150	B7
Crucible and lid (metal or ceramic)	15	B4
Deflagrating spoon	15	B4
Dropper bottle, 30 mL	90	B7, B14
Evaporating dish	15	B1
Forceps	15	B1, B10, B11, B12, B13
Gas-collecting bottle with rubber stopper, 250 mL	15	B4
Glass funnel	15	B1, B9
Glass stirring rod	15	B1, B4, B18, B19, B21
Glass tubing (10 cm)	15	B10
Glass vial with cap (for micro H_2 and O_2 generators)	60	B11, B12, B13
Glazed paper	30 sheets	B1
Gloves	30 pair	B1
Graduated cylinder, 100 mL	15	B1, B2
Graduated cylinder, 10 mL	15	B13, B16, B17

ChemFile Lab B

MASTER MATERIALS LIST continued

Equipment	Amount needed for 15 lab groups	Experiment
Graduated cylinder, 25 mL	15	B2, B18
Hot plate	15	B4
Mortar and pestle	15	B18
Petri dish with lid	30	B11, B12, B13, B21
pH test paper, narrow range	2 rolls	B9, B15
pH test paper, wide range	2 rolls	B15, B21
Piezoelectric sparking element	5	B13
Pipet bulb	15	B1
Pipet, graduated	15	B1
Pipet, thin-stemmed	1179	B3, B4, B5, B6, B7, B8, B9, B15, B16, B17, B19, B20, B21
Pliers	15	B21
Ring	15	B6
Ring stand	15	B1, B6
Rubber stopper (one-hole)	75	B1, B11, B12
Rubber stopper pegs	75	B11, B12
Rubber tubing	10.5 m	B1
Spatulas	15	B1, B7, B18
Spot plate	15	B7
Stopwatch	15	B18
Test tube (13 × 100)	15	B19
Test tube (large)	60	B1, B18
Test tube (small)	120	B9, B10, B11, B12
Test-tube clamp	15	B18
Test-tube rack	15	B18
Thermometer, Celsius, nonmercury, range −10 to 120°C	15	B2, B4, B18
Tongs	15	B21
Wash bottle	15	B1
Weighing paper	15 sheets	B1
Wire gauze (nonasbestos) with ceramic center	15	B1, B21
Wire stripper	1	B13
Wooden splints	60	B11, B12
Zinc strip	22.5 cm	B20

MASTER MATERIALS LIST continued

Miscellaneous Materials

Material	Amount needed for 15 lab groups	Experiment
Battery clips	15	B7
Battery, 9 V	15	B7
Candle	15	B6
Carbonated beverage	1 can	B4
Charcoal, activated	60 g	B7
Copper wire, 18 gauge	250 cm	B1, B20, B21
Drill, 1/8 inch	1	B11, B12, B13
Effervescent tablets	360	B18
Electrician's tape	1 roll	B7, B13
Film cannister	15	B7
Film cannister lid	20	B7, B13
Filter paper	135	B1, B9, B10
Fruit juice (3 kinds)	45 mL	B22
Heat-resistant gloves or pads	30 pair	B1
Heat-resistant mat	15	B1
Hook-up wire, 22 gauge, solid	450 cm	B7
Hot-glue gun	1	B13
Ice	3 L	B18
Index cards	15	B4
Iron nail	75	B7, B21
Iron wire	75 cm	B21
Labels	150	B7
Magnesium ribbon	160 cm	B20, B21
Metal samples (aluminum, copper, and lead)	270 g	B2
Paper towels	2 rolls	B7, B11
Piezoelectric fireplace butane lighter	5	B13
Plastic dishpan	2	B5
Plastic ruler, 15 cm	15	B2
Plastic wrap	1 roll	B8, B14
Platinum wire, No. 24	75 cm	B10
Refrigerator	1	B13

ChemFile Lab B

MASTER MATERIALS LIST continued

Material	Amount needed for 15 lab groups	Experiment
Sand	75 g	B1
Scissors	1	B13
Speaker hook-up wire, 24 AWG (Radio Shack #278-1059)	50 cm	B13
Steel wool	15 pads	B11
Sugar, white	60 g	B7
Tin can lid	15	B6
Toothpicks	150	B7
Tubing to cover LED nail	53 cm	B7
Wax	75 g	B7

Safety Equipment

Equipment	Amount needed for 15 lab groups	Experiment
Face shield	1	B3, B4, B8, B10, B12, B13, B15, B16, B17, B19, B20, B21
Impermeable gloves	1 pair	B3, B4, B8, B10, B12, B13, B15, B16, B17, B19, B20, B21
Lab apron	30	all
Safety goggles	30 pair	all

MASTER MATERIALS LIST continued

Chemical Inventory

There are many computer programs and data bases that you can use to track your chemical inventory. The following is an example of the information you could include within your inventory tracking system.

School Name: _____

Street Address: _____

City/County/Zip Code: _____

Chemical name and concentration or form	Estimated amount on hand	Amount and date purchased	Comments
Acetic acid, 5% solution	250 mL	2 L 9/97	Can substitute white vinegar
Acetic acid, glacial	900 mL	2 L 10/96	
$AgNO_3$, solid, technical grade	300 g	500 g 4/95	
$AgNO_3$, 1 M solution	200 mL	500 mL 9/97	Store in amber bottles
$Ba(NO_3)_2$, solid, technical grade	40 g	100 g 4/95	
$BaCl_2$, anhydrous form	130 g	250 g 4/95	
$BaCl_2 \cdot 5H_2O$, solid	15 g	100 g 4/95	Can substitute the anhydrous form

Incompatible Chemicals

The following listing should be considered when organizing and storing chemicals. Note that some chemicals on this list should no longer be in your lab due to their potential risks. Consult your state or local guidelines for more specific information on chemical hazards.

Chemical	Should not come in contact with
Acetic acid	Chromic acid, nitric acid, perchloric acid, ethylene glycol, hydroxyl compounds, peroxides, and permanganates
Acetone	Concentrated sulfuric acid and nitric acid mixtures
Acetylene	Silver, mercury and their compounds; bromine, chlorine, fluorine, and copper tubing
Alkali metals, powdered aluminum and magnesium	Water, carbon dioxide, carbon tetrachloride, and the halogens

ChemFile Lab B

MASTER MATERIALS LIST continued

Chemical	Should not come in contact with
Ammonia (anhydrous)	Mercury, hydrogen fluoride, and calcium hypochlorite
Ammonium nitrate (strong oxidizer)	Strong acids, metal powders, chlorates, sulfur, flammable liquids, and finely-divided organic materials
Aniline	Nitric acid and hydrogen peroxide
Bromine	Ammonia, acetylene, butane, hydrogen, sodium carbide, turpentine, and finely-divided metals
Carbon (activated)	Calcium hypochlorite, all oxidizing agents
Chlorates	Ammonium salts, strong acids, powdered metals, sulfur, and finely-divided organic materials
Chromic acid	Glacial acetic acid, camphor, glycerin, naphthalene, turpentine, low-molar mass alcohols, and flammable liquids
Chlorine	Same as bromine
Copper	Acetylene and hydrogen peroxide
Flammable liquids	Ammonium nitrate, chromic acid, hydrogen peroxide, sodium peroxide, nitric acid, and the halogens
Hydrocarbons (butane, propane, gasoline, turpentine)	Fluorine, chlorine, bromine, chromic acid, and sodium peroxide
Hydrofluoric acid	Ammonia
Hydrogen peroxide	Most metals and their salts, flammable liquids, and other combustible materials
Hydrogen sulfide	Nitric acid and certain other oxidizing gases
Iodine	Acetylene and ammonia
Nitric acid	Glacial acetic acid, chromic and hydrocyanic acids, hydrogen sulfide, flammable liquids, and flammable gases that are easily nitrated
Oxygen	Oils, grease, hydrogen, flammable substances
Perchloric acid	Acetic anhydride, bismuth and its alloys, alcohols, paper, wood, and other organic materials
Phosphorus pentoxide	Water
Potassium permanganate	Glycerin, ethylene glycol, and sulfuric acid
Silver	Acetylene, ammonium compounds, oxalic acid, and tartaric acid
Sodium peroxide	Glacial acetic acid, acetic anhydride, methanol, carbon disulfide, glycerin, benzaldehyde, and water
Sulfuric acid	Chlorates, perchlorates, permanganates, and water

Microscale Experiments

HOLT, RINEHART AND WINSTON
Harcourt Brace & Company
Austin • New York • Orlando • Atlanta • San Francisco • Boston • Dallas • Toronto • London

Copyright © by Holt, Rinehart and Winston
All rights reserved. No part of this publication may be reproduced or transmitted in any form or by any means, electronic or mechanical, including photocopy, recording, or any information storage and retrieval system, without permission in writing from the publisher.

Teachers using HOLT CHEMFILE may photocopy Blackline Masters in complete pages in sufficient quantities for classroom use only and not for resale.

Cover: HRW Photo by Sam Dudgeon

Printed in the United States of America

ISBN 0-03-051924-1

8 9 10 11 12 13 14 15   022   05 04 03 02 01 00

LAB MANUAL B CONTENTS

	Introduction	iv
	Safety	vi
	Labeling of Chemicals	xi
	Laboratory Techniques	xii
B1	Laboratory Procedures	1
B2	Accuracy and Precision: Calibrating a Pipet	21
B3	Relative Solubility of Transition Elements	29
B4	Periodicity of Properties of Oxides	33
B5	Reactivity of Halide Ions	39
B6	Chemical Bonds	43
B7	Conductivity as an Indicator of Bond Type	47
B8	Test for Iron(II) and Iron(III)	53
B9	Colored Precipitates	57
B10	Simple Qualitative Analysis	63
B11	Generating and Collecting O_2	67
B12	Generating and Collecting H_2	73
B13	Testing Reaction Combinations of H_2 and O_2	79
B14	Reacting Ionic Species in Aqueous Solution	87
B15	Hydronium Ion Concentration, pH	91
B16	Titration of an Acid with a Base	95
B17	Percentage of Acetic Acid in Vinegar	99
B18	Clock Reactions	105
B19	Equilibrium	109
B20	Oxidation-Reduction Reactions	115
B21	Cathodic Protection: Factors That Affect the Corrosion of Iron	119
B22	Determination of Vitamin C in Fruit Juices	125

ChemFile Lab B

Introduction to the Lab Program

STRUCTURE OF THE EXPERIMENTS

INTRODUCTION
The opening paragraphs set the theme for the experiment and summarize its major concepts.

OBJECTIVES
Objectives highlight the key concepts to be learned in the experiment and emphasize the science process skills and techniques of scientific inquiry.

MATERIALS
These lists enable you to organize all apparatus and materials needed to perform the experiment. Knowing the concentrations of solutions is vital. You often need this information to perform calculations and to answer the questions at the end of the experiment.

SAFETY
Safety cautions are placed at the beginning of the experiment to alert you to procedures that may require special care. Before you begin, you should review with the safety issues that apply to the experiment.

PROCEDURE
By following the procedures of an experiment, you are performing concrete laboratory operations that duplicate the fact-gathering techniques used by professional chemists. You are learning skills in the laboratory. The procedures tell you how and where to record observations and data.

DATA AND CALCULATIONS TABLES
The data that you will collect during each experiment should be recorded in the labeled Data Tables provided. The entries you make in a Calculations Table emphasize the mathematical, physical, and chemical relationships that exist among the accumulated data. Both tables should help them to think logically and to formulate their conclusions about what occurred during the experiment.

CALCULATIONS
Space is provided for all computations based on the data you have gathered. Write your final answers on the spaces provided or circle the final answer to a calculation.

QUESTIONS
Based on the data and calculations, you should be able to develop a plausible explanation for the phenomena observed during the experiment. Specific questions are asked that require you to draw on the concepts you have learned.

GENERAL CONCLUSIONS
This section asks broader questions that bring together the results and conclusions of the experiment and relate them to other situations.

Safety in the Chemistry Laboratory

Chemicals are not toys

Any chemical can be dangerous if it is misused. Always follow the instructions for the experiment. Pay close attention to the safety notes. Do not do anything differently unless told to do so by your teacher.

Chemicals, even water, can cause harm. The trick is to know how to use chemicals correctly so that they will not cause harm. If you follow the rules stated in the following pages, pay attention to your teacher's directions, and follow the cautions on chemical labels and the experiments, then you will be using chemicals correctly.

These safety rules always apply in the lab

1. **Always wear a lab apron and safety goggles.**
   Even if you aren't working on an experiment, laboratories contain chemicals that can damage your clothing, so wear your apron and keep the strings of the apron tied. Because chemicals can cause eye damage, even blindness, you must wear safety goggles. If your safety goggles are uncomfortable or get clouded up, ask your teacher for help. Try lengthening the strap a bit, washing the goggles with soap and warm water, or using an antifog spray.

2. **No contact lenses are allowed in the lab.**
   Even while wearing safety goggles, chemicals could get between contact lenses and your eyes and cause irreparable eye damage. If your doctor requires that you wear contact lenses instead of glasses, then you should wear eye-cup safety goggles in the lab. Ask your doctor or your teacher how to use this very important and special eye protection.

3. **Never work alone in the laboratory.**
   You should always do lab work only under the supervision of your teacher.

4. **Wear the right clothing for lab work.**
   Necklaces, neckties, dangling jewelry, long hair, and loose clothing can cause you to knock things over or catch items on fire. Tuck in neckties or take them off. Do not wear a necklace or other dangling jewelry, including hanging earrings. It isn't necessary, but it might be a good idea to remove your wristwatch so that it is not damaged by a chemical splash.

   Pull back long hair, and tie it in place. Nylon and polyester fabrics burn and melt more readily than cotton, so wear cotton clothing if you can. It's best to wear fitted garments, but if your clothing is loose or baggy, tuck it in or tie it back so that it does not get in the way or catch on fire.

   Wear shoes that will protect your feet from chemical spills — no open-toed shoes or sandals and no shoes with woven leather straps. Shoes made of solid leather or a polymer are much better than shoes made of cloth. Also, wear pants, not shorts or skirts.

5. **Only books and notebooks needed for the experiment should be in the lab.**
   Do not bring other textbooks, purses, bookbags, backpacks, or other items into the lab; keep these things in your desk or locker.

ChemFile LAB B

SAFETY IN THE CHEMISTRY LABORATORY continued

6. **Read the entire experiment before entering the lab.**
   Memorize the safety precautions. Be familiar with the instructions for the experiment. Only materials and equipment authorized by your teacher should be used. When you do the lab work, follow the instructions and the safety precautions described in the directions for the experiment.

7. **Read chemical labels.**
   Follow the instructions and safety precautions stated on the labels. Know the location of Materials Safety Data Sheets for chemicals

8. **Walk carefully in the lab.**
   Sometimes you will carry chemicals from the supply station to your lab station. Avoid bumping other students and spilling the chemicals. Stay at your lab station at other times.

9. **Food, beverages, chewing gum, cosmetics, and smoking are NEVER allowed in the lab.**
   You already know this.

10. **Never taste chemicals or touch them with your bare hands.**
    Also, keep your hands away from your face and mouth while working, even if you are wearing gloves.

11. **Use a sparker to light a Bunsen burner.**
    Do not use matches. Be sure that all gas valves are turned off and that all hot plates are turned off and unplugged when you leave the lab.

12. **Be careful with hot plates, Bunsen burners, and other heat sources.**
    Keep your body and clothing away from flames. Do not touch a hot plate after it has just been turned off. It is probably hotter than you think. The same is true of glassware, crucibles, and other things after you remove them from a hot plate, drying oven, or the flame of a Bunsen burner.

13. **Do not use electrical equipment with frayed or twisted cords or wires.**

14. **Be sure your hands are dry before using electrical equipment**
    Before plugging an electrical cord into a socket, be sure the electrical equipment is turned off. When you are finished with it, turn it off. Before you leave the lab, unplug it, but be sure to turn it off first.

15. **Do not let electrical cords dangle from work stations; dangling cords can cause tripping or electrical shocks.**
    The area under and around electrical equipment should be dry; cords should not lie in puddles of spilled liquid.

16. **Know fire drill procedures and the locations of exits.**

17. **Know the location and operation of safety showers and eyewash stations.**

18. **If your clothes catch on fire, walk to the safety shower, stand under it, and turn it on.**

19. **If you get a chemical in your eyes, walk immediately to the eyewash station, turn it on, and lower your head so that your eyes are in the running water.**
    Hold your eyelids open with your thumbs and fingers, and roll your eyeballs around. You have to flush your eyes continuously for at least 15 min. Call your teacher while you are doing this.

SAFETY IN THE CHEMISTRY LABORATORY continued

20. **If you have a spill on the floor or lab bench, call your teacher rather than trying to clean it up by yourself.**
    Your teacher will tell you if it is OK for you to do the cleanup; if it is not, your teacher will know how the spill should be cleaned up safely.

21. **If you spill a chemical on your skin, wash it off under the sink faucet, and call your teacher.**
    If you spill a solid chemical on your clothing, brush it off carefully so that you do not scatter it, and call your teacher. If you get a liquid on your clothing, wash it off right away if you can get it under the sink faucet, and call your teacher. If the spill is on clothing that will not fit under the sink faucet, use the safety shower. Remove the affected clothing while under the shower, and call your teacher. (It may be temporarily embarrassing to remove your clothing in front of your class, but failing to flush that chemical off your skin could cause permanent damage.)

22. **The best way to prevent an accident is to stop it before it happens.**
    If you have a close call, tell your teacher so that you and your teacher can find a way to prevent it from happening again. Otherwise, the next time, it could be a harmful accident instead of just a close call.

23. **All accidents should be reported to your teacher, no matter how minor.**
    Also, if you get a headache, feel sick to your stomach, or feel dizzy, tell your teacher immediately.

24. **For all chemicals, take only what you need.**
    On the other hand, if you do happen to take too much and have some left over, DO NOT put it back in the bottle. If somebody accidentally puts a chemical into the wrong bottle, the next person to use it will have a contaminated sample. Ask your teacher what to do with any leftover chemicals.

25. **NEVER take any chemicals out of the lab.**
    You should already know this rule.

26. **Horseplay and fooling around in the lab are very dangerous.**
    NEVER be a clown in the laboratory.

27. **Keep your work area clean and tidy.**
    After your work is done, clean your work area and all equipment.

28. **Always wash your hands with soap and water before you leave the lab.**

29. **Whether or not the lab instructions remind you, ALL of these rules APPLY ALL OF THE TIME.**

QUIZ

Determine which safety rules apply to the following.

- Tie back long hair, and confine loose clothing. (Rule ? applies.)
- Never reach across an open flame. (Rule ? applies.)
- Use proper procedures when lighting Bunsen burners. Turn off hot plates, Bunsen burners, and other heat sources when not in use. (Rule ? applies.)
- Heat flasks and beakers on a ring stand with wire gauze between the glass and the flame. (Rule ? applies.)
- Use tongs when heating containers. Never hold or touch containers with your hands while heating them. Always allow heated materials to cool before handling them. (Rule ? applies.)
- Turn off gas valves when not in use. (Rule ? applies.)

SAFETY IN THE CHEMISTRY LABORATORY continued

SAFETY SYMBOLS

To highlight specific types of precautions, the following symbols are used in the experiments. Remember that no matter what safety symbols and instructions appear in each experiment, all of the 29 safety rules described previously should be followed at all times.

Eye and clothing protection

- Wear laboratory aprons in the laboratory. Keep the apron strings tied so that they do not dangle.
- Wear safety goggles in the laboratory at all times. Know how to use the eyewash station.

Chemical safety

- Never taste, eat, or swallow any chemicals in the laboratory. Do not eat or drink any food from laboratory containers. Beakers are not cups, and evaporating dishes are not bowls.
- Never return unused chemicals to the original container.
- Some chemicals are harmful to the environment. You can help protect the environment by following the instructions for proper disposal.
- It helps to label the beakers and test tubes containing chemicals.
- Never transfer substances by sucking on a pipette or straw; use a suction bulb.
- Never place glassware, containers of chemicals, or anything else near the edges of a lab bench or table.

Caustic substances

- If a chemical gets on your skin or clothing or in your eyes, rinse it immediately, and alert your teacher.
- If a chemical is spilled on the floor or lab bench, tell your teacher, but do not clean it up yourself unless your teacher says it is OK to do so.

Heating safety

- When heating a chemical in a test tube, always point the open end of the test tube away from yourself and other people. (This is another new rule.)

Explosive precaution

- Use flammable liquids only in small amounts.
- When working with flammable liquids, be sure that no one else in the lab is using a lit Bunsen burner or plans to use one. Make sure there are no other heat sources present.

SAFETY IN THE CHEMISTRY LABORATORY continued

Hand safety

- Always wear gloves or cloths to protect your hands when cutting, fire polishing, or bending hot glass tubing. Keep cloths clear of any flames.
- Never force glass tubing into rubber tubing, rubber stoppers, or corks. To protect your hands, wear heavy leather gloves or wrap toweling around the glass and the tubing, stopper, or cork, and gently push in the glass tubing.
- Use tongs when heating test tubes. Never hold a test tube in your hand to heat it.
- Always allow hot glasswear to cool before handling.

Glassware safety

- Check the condition of glassware before and after using it. Inform your teacher of any broken, chipped, or cracked glassware because it should not be used.
- Do not pick up broken glass with your bare hands. Place broken glass in a specially designated disposal container.

Gas precaution

- Do not inhale fumes directly. When instructed to smell a substance, use your hand, wave the fumes toward your nose, and inhale gently. (Some people say "waft the fumes.")

Radiation precaution

- Always wear gloves when handling a radioactive source.
- Always wear safety goggles when performing experiments with radioactive materials.
- Always wash your hands and arms thoroughly after working with radioactive materials.

Hygiene precaution

- Keep your hands away from your face and mouth.
- Always wash your hands before leaving the laboratory.

Any time you see any of the safety symbols you should remember that all 29 of the numbered laboratory rules always apply.

Labeling of Chemicals

In any science laboratory the *labeling* of chemical containers, reagent bottles, and equipment is essential for safe operations. Proper labeling can lower the potential for accidents that occur as a result of misuse. Labels and equipment instructions should be read several times before using. Be sure that you are using the correct items, that you know how to use them, and that you are aware of any hazards or precautions associated with their use.

All chemical containers and reagent bottles should be labeled prominently and accurately using labeling materials that are not affected chemicals.

Chemical labels should contain the following information.

1. **Name of chemical and the chemical formula**
2. **Statement of possible hazards** This is indicated by the use of an appropriate signal word, such as DANGER, WARNING, or CAUTION. This signal word usually is accompanied by a word that indicates the type of hazard present such as POISON, CAUSES BURNS, EXPLOSIVE or FLAMMABLE. Note that this labeling should not take the place of reading the appropriate Material Safety Data Sheet for a chemical.
3. **Precautionary measures** Precautionary measures describe how users can avoid injury from the hazards listed on the label. Examples include: "Use only with adequate ventilation," and "Do not get in eyes or on skin or clothing."
4. **Instructions in case of contact or exposure** If accidental contact or exposure does occur immediate treatment is often necessary to minimize injury. Such treatment usually consists of proper first-aid measures that can be used before a physician administers treatment. An example is: "In case of contact, flush with large amounts of water; for eyes, rinse freely with water for 15 minutes and get medical attention immediately"
5. **The date of preparation and the name of the person who prepared the chemical** This information is important for maintaining a safe chemical inventory.

Suggested Labeling Scheme

Name of contents	Hydrochloric Acid	
	6 M HCl	Chemical formula and concentration or physical state
Statements of possible hazards and precautionary measures	WARNING! CAUSTIC and CORROSIVE-CAUSES BURNS CAUTION! Avoid contact with skin and eyes. Avoid breathing vapors.	
	IN CASE OF CONTACT: Immediately flush skin or eyes with large amounts of water for at least 15 minutes; for eyes, get medical attention immediately!	Hazard Instructions for contact or overexposure
Date prepared or obtained	May 8, 1989 Prepared by Betsy Byron Faribault High School, Faribault, Minnesota	Manufacturer (Commercially obtained) or preparer (Locally made)

Laboratory Techniques

| FIGURE A | FIGURE B | FIGURE C |

DECANTING AND TRANSFERRING LIQUIDS

1. The safest way to transfer a liquid from a graduated cylinder to a test tube is shown in Figure A. The liquid is transferred at arm's length with the elbows slightly bent. This position enables you to see what you are doing and still maintain steady control.
2. Sometimes liquids contain particles of insoluble solids that sink to the bottom of a test tube or beaker. Use one of the methods shown below to separate a supernatant (the clear fluid) from insoluble solids.
   a. Figure B shows the proper method of decanting a supernatant liquid in a test tube.
   b. Figure C shows the proper method of decanting a supernatant liquid in a beaker by using a stirring rod. The rod should touch the wall of the receiving container. Hold the stirring rod against the lip of the beaker containing the supernatant liquid. As you pour, the liquid will run down the rod and fall into the beaker resting below. Using this method, the liquid will not run down the side of the beaker from which you are pouring.

HEATING SUBSTANCES AND EVAPORATING SOLUTIONS

1. Use care in selecting glassware for high-temperature heating. The glassware should be heat resistant.
2. When heating glassware using a gas flame, use a ceramic-centered wire gauze to protect glassware from direct contact with the flame. Wire gauzes can withstand extremely high temperatures and will help prevent glassware from breaking. Figure D shows the proper setup for evaporating a solution over a water bath.

FIGURE D

ChemFile LAB B

LABORATORY TECHNIQUES continued

FIGURE E

FIGURE F

3. In some experiments you are required to heat a substance to high temperatures in a porcelain crucible. Figure E shows the proper apparatus setup used to accomplish this task.
4. Figure F shows the proper setup for evaporating a solution in a porcelain evaporating dish with a watch glass cover that prevents spattering.
5. Glassware, porcelain, and iron rings that have been heated may *look* cool after they are removed from a heat source, but these items can still burn your skin even after several minutes of cooling. Use tongs, test-tube holders, or heat-resistant mitts and pads whenever you handle this apparatus.
6. You can test the temperature of questionable beakers, ring stands, wire gauzes, or other pieces of apparatus that have been heated by holding the back of your hand close to their surfaces before grasping them. You will be able to feel any heat generated from the hot surfaces. DO NOT TOUCH THE APPARATUS. Allow plenty of time for the apparatus to cool before handling.

HOW TO POUR LIQUID FROM A REAGENT BOTTLE

1. Read the label at least three times before using the contents of a reagent bottle.
2. Never lay the stopper of a reagent bottle on the lab table.
3. When pouring a caustic or corrosive liquid into a beaker use a stirring rod to avoid drips and spills. Hold the stirring rod against the lip of the reagent bottle. Estimate the amount of liquid you need and pour this amount along the rod into the beaker. See Figure G.

FIGURE G

LABORATORY TECHNIQUES continued

4. Extra precaution should be taken when handling a bottle of acid. Remember the following important rules: Never add water to any concentrated acid, particularly sulfuric acid, because the mixture can splash and will generate a lot of heat. To dilute any acid, add the acid to water in small quantities, while stirring slowly. Remember the "triple A's"-Always Add Acid to water.
5. Examine the outside of the reagent bottle for any liquid that has dripped down the bottle or spilled on the counter top. Your teacher will show you the proper procedures for cleaning up a chemical spill.
6. Never pour reagents back into stock bottles. At the end of the experiment, your teacher will tell you how to dispose of any excess chemicals.

HOW TO HEAT MATERIAL IN A TEST TUBE

1. Check to see that the test tube is heat-resistant.
2. Always use a test tube holder or clamp when heating a test tube.
3. Never point a heated test tube at anyone, because the liquid may splash out of the test tube.
4. Never look down into the test tube while heating it.
5. Heat the test tube from the upper portions of the tube downward and continuously move the test tube as shown in Figure H. Do not heat any one spot on the test tube. Otherwise a pressure build-up may cause the bottom of the tube to blow out.

HOW TO USE A MORTAR AND PESTLE

1. A mortar and pestle should be used for grinding only one substance at a time. See Figure I.
2. Never use a mortar and pestle for simultaneously mixing different substances.
3. Place the substance to be broken up into the mortar
4. Pound the substance with the pestle and grind to pulverize.
5. Remove the powdered substance with a porcelain spoon.

DETECTING ODORS SAFELY

1. Test for the odor of gases by wafting your hand over the test tube and cautiously sniffing the fumes as shown in Figure J.
2. Do not inhale any fumes directly.
3. Use a fume hood whenever poisonous or irritating fumes are evolved. DO NOT waft and sniff poisonous or irritating fumes.

FIGURE H   **FIGURE I**   **FIGURE J**

ChemFile LAB B

Name _____

Date _____ Class _____

EXPERIMENT B1

Laboratory Procedures

OBJECTIVES

Recommended time:
2 lab periods

- **Observe** proper safety techniques with all laboratory equipment.
- **Use** laboratory apparatus skillfully and efficiently.
- **Recognize** the names and functions of all apparatus in the laboratory.
- **Develop** a positive approach toward laboratory safety.

INTRODUCTION

The best way to become familiar with chemical apparatus is to handle the pieces yourself in the laboratory. This experiment is divided into several parts in which you will learn how to adjust the gas burner, insert glass tubing into a rubber stopper, use a balance, handle solids, measure liquids, filter a mixture, and measure temperature and heat. Great emphasis is placed on safety precautions that should be observed whenever you perform an experiment and use the apparatus. Several useful manipulative techniques are also illustrated on pages xi through xiii. In many of the later experiments, references will be made to these "Laboratory Techniques." In later experiments you will also be referred to the safety precautions and procedures explained in all parts of this experiment. It is important that you develop a positive approach to a safe and healthful environment in the lab.

SAFETY

Required Precautions

- Discuss all safety symbols and caution statements with students.
- Safety goggles and a lab apron must be worn at all times.
- Read all safety cautions and discuss them with your students.
- Loose hair and clothing must be tied back.
- If students find a gas valve that has not been turned off properly, the room must be thoroughly ventilated, using the hood and windows in order to prevent a fire hazard.

  Always wear safety goggles and a lab apron to protect your eyes and clothing. If you get a chemical in your eyes, immediately flush the chemical out at the eyewash station while calling to your teacher. Know the location of the emergency lab shower and eyewash station and the procedure for using them.

 Do not touch any chemicals used in the laboratory. If you get a chemical on your skin or clothing, wash the chemical off at the sink while calling to your teacher. Make sure you carefully read the labels and follow the precautions on all containers of chemicals that you use. If there are no precautions stated on the label, ask your teacher what precautions to follow. Never return leftover chemicals to their original containers; take only small amounts to avoid wasting supplies.

 Never put broken glass into a regular waste container. Broken glass should be disposed of separately according to your teacher's instructions.

ChemFile

EXPERIMENT B1 **1**

EXPERIMENT B1 continued

 When using a Bunsen burner, confine long hair and loose clothing. If your clothing catches on fire, WALK to the emergency lab shower, and use it to put out the fire. Do not heat glassware that is broken, chipped, or cracked. Use tongs or a hot mitt to handle heated glassware and other equipment because heated glassware does not look hot.

When you insert glass tubing into stoppers, lubricate the glass with water or glycerin and protect your hands and fingers: Wear leather gloves or place folded cloth pads between both of your hands and the glass tubing. Then *gently* push the tubing into the stopper hole. In the same way, protect your hands and fingers when removing glass tubing from stoppers and from rubber or plastic tubing.

MATERIALS

PART 1 THE BURNER

- Bunsen burner and related equipment
- copper wire, 18 gauge
- evaporating dish
- forceps
- heat-resistant mat
- sparker

PROCEDURE

Procedural Tips

• Although all the techniques are carefully described, there is no substitute for seeing each of them demonstrated before students try them themselves.

1. The Bunsen burner is commonly used as a source of heat in the laboratory. Look at Figure A as you examine your Bunsen burner and identify the parts. Although the details of construction vary among burners, each has a gas inlet located in the base, a vertical tube or barrel in which the gas is mixed with air, and adjustable openings or ports in the base of the barrel. These ports admit air to the gas stream. The burner may have an adjustable needle valve to regulate the flow of gas. In some models the gas flow is regulated simply by adjusting the gas valve on the supply line. The burner is always turned off at the gas valve, never at the needle valve.

FIGURE A

EXPERIMENT B1 continued

Margin notes (left column):

- Close supervision is necessary if students work with glass tubing. Emphasize that they should never attempt to force tubing into stoppers with excessive force. Advise them to use a lubricant such as water or glycerin to help ease the tubing in.

- Stress to students that hot tubing or other glassware does not look hot, and care should be taken to avoid burns.

- Advise students to read and follow the procedures carefully.

Disposal

- There are no special disposal guidelines. The sodium chloride can be reused by each class if you designate a special container for NaCl disposal after each part. Keep this NaCl separate from the NaCl you use as a reagent. The same is true for the copper wire and the sand.

CAUTION Before you light the burner, check to see that you and your partner have taken the following safety precautions against fires: Wear safety goggles, aprons, and gloves. Confine long hair and loose clothing: tie long hair at the back of the head and away from the front of the face, and roll up long sleeves on shirts, blouses, and sweaters away from the wrists. You should also know the locations of fire extinguishers, fire blankets, safety showers, and sand buckets and the procedure for using them in case of a fire.

2. When lighting the burner, partially close the ports at the base of the barrel, turn the gas full on, hold the sparker about 5 cm above the top of the burner, and proceed to light. The gas flow may then be regulated by adjusting the gas valve until the flame has the desired height. If a very low flame is needed, remember that the ports should be partially closed when the gas pressure is reduced. Otherwise the flame may burn inside the base of the barrel. When the flame is improperly burning in this way, the barrel will get very hot, and the flame will produce a poisonous gas, carbon monoxide.

CAUTION If the flame is burning inside the base of the barrel, immediately turn off the gas at the gas valve. Do not touch the barrel, for it is extremely hot. Allow the barrel of the burner to cool and then proceed as follows:

Begin again, but first decrease the amount of air admitted to the burner by partially closing the ports. Turn the gas full on and then relight the burner. Control the height of the flame by adjusting the gas valve. By taking these steps, you should acquire a flame that is burning safely and is easily regulated.

3. Once you have a flame that is burning safely and steadily, you can experiment by completely closing the ports at the base of the burner. What effect does this have on the flame?

When the ports at the base are closed, a yellow flame results.

Using the forceps, hold an evaporating dish in the tip of the flame for about 3 min. Place the dish on a heat-resistant mat and allow the dish to cool. Then examine the bottom of the dish. Describe the results and suggest a possible explanation.

The yellow flame leaves a black deposit on the bottom of the evaporating dish. The lack of air to the burner results in incomplete combustion, the luminous flame, and the deposit of soot.

Such a flame is seldom used in the lab. For laboratory work, you should adjust the burner so that the flame is free of yellow color, nonluminous, and also free of the roaring sound caused by admitting too much air.

4. Regulate the flow of gas so that the flame extends roughly 8 cm above the barrel. Now adjust the supply of air until you have a quiet, steady flame with

EXPERIMENT B1 continued

a sharply defined, light blue inner cone. This adjustment gives the highest temperature possible with your burner. Using the forceps, insert a 10 cm piece of copper wire into the flame just above the barrel. Lift the wire slowly up through the flame. Where is the hottest portion of the flame located?

The hottest portion of the nonluminous flame is just above the tip of the light blue cone.

Hold the wire in this part of the flame for a few seconds. What happens?

The copper wire becomes red hot and begins to soften (melt).

5. Shut off the gas burner. Now think about what you have just observed in steps **3** and **4**. Why is the nonluminous flame preferred over the yellow luminous flame in the laboratory?

The nonluminous flame burns cleaner than the yellow flame.

6. Clean the evaporating dish and put away the burner. All the equipment you store in the lab locker or drawer should be completely cool, clean, and dry. Be sure that the valve on the gas jet is completely shut off. Remember to wash your hands thoroughly with soap at the end of each laboratory period.

MATERIALS

PART 2 GLASS MANIPULATIONS

- cloth pads or leather gloves
- glass funnel
- rubber hose
- rubber stopper, 1-hole
- water or glycerin

PROCEDURE

1. Inserting glass tubing into rubber stoppers can be very dangerous. The following precautions should be observed to prevent injuries:
   a. Never attempt to insert glass tubing that has a jagged end. All glass tubing should be fire polished before it is inserted into a rubber stopper.
   b. Use water or glycerin as a lubricant on the end of the glass tubing before inserting it into a rubber stopper. Ask your teacher for the proper lubricant. **CAUTION Protect your hands and fingers when inserting glass tubing into a rubber stopper.**
   c. Wear leather gloves or place folded cloth pads between your hands and the glass tubing. Hold the glass tubing as close as possible to the part where it is to enter the rubber stopper. Always point the glass tubing away from the palm of your hand that holds the stopper, as shown in Figure B on the next page. Using a twisting motion, gently push the tubing into the stopper hole.
   d. At the end of the experiment, put on leather gloves or place folded cloth pads between your hands and the glass tubing and remove the rubber stoppers from the tubing to keep them from sticking or "freezing" to the glass. Use a lubricant as directed in step **1b** if the stopper or tubing won't budge.

EXPERIMENT B1 continued

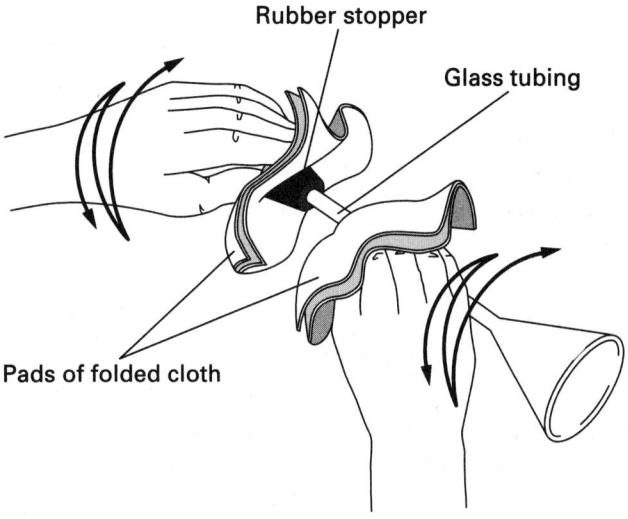

FIGURE B

2. When inserting glass tubing into a rubber or plastic hose, observe the same precautions discussed in steps **1a–1c.** The glass tubing should be lubricated before insertion into the rubber or plastic hose. The rubber hose should be cut at an angle before the insertion of the glass tubing. The angled cut in the hose allows the rubber to stretch more readily.
CAUTION Protect your hands when inserting or removing glass tubing.
   At the end of an experiment, immediately remove the glass tubing from the hose. When disassembling, follow the precautions that were given in step **1d.**
   Carefully follow these precautions and techniques whenever an experiment requires that you insert glass tubing into either a rubber stopper or a rubber or plastic hose. You will be referred to these safety precautions throughout the lab course.

MATERIALS

PART 3 HANDLING SOLIDS
- glazed paper
- sodium chloride
- spatula
- test tube

PROCEDURE

1. Solids are usually kept in wide-mouthed bottles. A spatula should be used to dip out the solid as shown in Figure C.

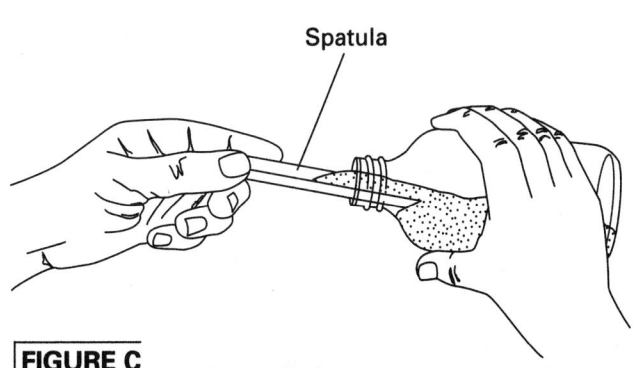

FIGURE C

EXPERIMENT B1 continued

FIGURE D

CAUTION Do not touch chemicals with your hands. Some chemical reagents readily pass through the skin into the bloodstream and can cause serious health problems. Some chemicals are corrosive. Always wear an apron and safety goggles when handling chemicals. Carefully check the label on the reagent bottle or container before removing any of the contents. Never use more of a chemical than directed. You should also know the locations of the emergency lab shower and eyewash station and the procedure for using them in case of an accident.

Using a spatula remove a quantity of sodium chloride from its reagent bottle. In order to transfer the sodium chloride to a test tube, first place it on a piece of glazed paper about 10 cm square. Roll the paper into a cylinder and slide it into a test tube that is lying flat on the table. When you lift the tube to a vertical position and tap the paper gently, the solid will slide down into the test tube, as shown in Figure D.

CAUTION Never try to pour a solid from a bottle into a test tube. As a precaution against contamination, never pour unused chemicals back into their reagent bottles.

2. Dispose of the solid sodium chloride and glazed paper in the waste jars or containers provided by your teacher.
CAUTION Never discard chemicals or broken glassware in the wastepaper basket. This is an important safety precaution against fires, and it prevents personal injuries (such as hand cuts) to anyone who empties the wastepaper basket.

 3. Remember to clean up the lab station and wash your hands at the end of this part of the experiment.

MATERIALS

PART 4 THE BALANCE
- balance, centigram
- glazed paper
- sodium chloride
- spatula
- weighing paper

EXPERIMENT B1 continued

FIGURE E

PROCEDURE

1. When a balance is required for determining mass, you will use a centigram balance like the one shown in Figure E. The centigram balance has a readability of 0.01 g. This means that your mass readings should all be recorded to the nearest 0.01 g.

2. Before using the balance, always check to see if the pointer is resting at zero. If the pointer is not at zero, check the riders on the scales. If all the scale riders are at zero, turn the zero adjust knob until the pointer rests at zero. The zero adjust knob is usually located at the far left end of the balance beam as shown in Figure E. Note: The balance will not adjust to zero if the movable pan has been removed. **Never place chemicals or hot objects directly on the balance pan.** Always use weighing paper or a glass container. Chemicals can permanently damage the surface of the balance pan and affect the accuracy of measurements.

3. In many experiments you will be asked to determine the mass of a specified amount of a chemical solid. Use the following procedure to obtain approximately 13 grams of sodium chloride.
   a. Make sure the pointer on the balance is set at zero. Obtain a piece of weighing paper and place it on the balance pan. Determine the mass of the paper by adjusting the riders on the various scales. Record the mass of the weighing paper to the nearest 0.01 g.

   Mass of paper: Student answer

   b. Add 13 grams to the balance by sliding the rider on the 100 g scale to 10 and the rider on the 10 g scale to 3.
   c. Using a spatula, obtain a quantity of sodium chloride from the reagent bottle and place it on a separate piece of glazed paper.

EXPERIMENT B1 continued

   d. Now slowly pour the sodium chloride from the glazed paper onto the weighing paper on the balance pan, until the pointer once again comes to zero. In most cases, you will only have to be close to the specified mass. Do not waste time trying to obtain exactly 13.00 g. Instead, read the exact mass when the pointer rests close to zero and you have around 13 g of sodium chloride in the balance pan. The mass might be 13.18 g. Record your exact mass of sodium chloride and the weighing paper to the nearest 0.01 g. (Hint: Remember to subtract the mass of the weighing paper to find the mass of sodium chloride.)

   Mass of NaCl and paper: Student answer

4. Wash your hands thoroughly with soap and water at the end of each lab period.

MATERIALS

PART 5 MEASURING LIQUIDS

- 50 mL beaker
- 250 mL beaker
- 100 mL graduated cylinder
- buret
- buret clamp
- funnel
- pipet
- ring stand
- water

PROCEDURE

1. For approximate measurements of liquids, a graduated cylinder such as the one shown in Figure F is generally used. These cylinders are usually graduated in milliliters (mL). They may also have a second column of graduations reading from top to bottom. Examine your cylinder for these markings. Record the capacity and describe the scale of your cylinder in the space below.

   Observation:

   _____

   _____

   _____

2. A pipet or a buret is used for more accurate volume measurements. Pipets are made in many sizes and are used to deliver measured volumes of liquids. A pipet is fitted with a suction bulb as shown in Figure G on the next page. The bulb is used to withdraw air from the pipet while drawing up the liquid to be measured. **CAUTION Always use the suction bulb. NEVER pipet by mouth.**

FIGURE F

EXPERIMENT B1 continued

FIGURE G

3. Burets are used for delivering any desired quantity of liquid up to the capacity of the buret. Many burets are graduated in tenths of milliliters. When using a buret, follow these steps:
   a. Clamp the buret in position on a ring stand as shown in Figure H.
   b. Place a 250 mL beaker under the tip of the buret. The beaker serves to catch any liquid that is released.
   c. Pour a quantity of liquid you want to measure from the liquid's reagent bottle into a 50 mL beaker. (NOTE: In this first trial you will use water.) Use a glass funnel in the top of the buret to avoid spills when pouring the liquid from the beaker. Carefully check the label of the reagent bottle before removing any liquid. **CAUTION Never pour a liquid directly from its reagent bottle into the buret. You should first pour the liquid into a small, clean, and dry beaker (50 mL) that is easy to handle. Then pour the liquid from the small beaker into the buret. This simple method will prevent unnecessary spillage. Never pour any unused liquid back into the reagent bottle.**
   d. Fill the buret with the liquid and then open the stopcock to release enough liquid to fill the tip below the stopcock and bring the level of the liquid within the scale. The height at which the liquid stands is then read accurately. Practice this procedure several times by pouring water into the buret and emptying it through the stopcock.

FIGURE H

4. Notice that the surface of a liquid in the buret is slightly curved. It is concave if it wets the glass and convex if it does not wet the glass. Such a curved surface is called a meniscus. If a liquid wets the glass, read the bottom of the meniscus, as shown in Figure I on the next page. This is the line CB. If you read the markings at the top of the meniscus, AD, you will get an incorrect reading. Locate the bottom of the meniscus and read the water level in your buret.

EXPERIMENT B1 continued

FIGURE I

Buret reading: <u>Student answer</u>

5. After you have taken your first buret reading, open the stopcock to release some of the liquid. Then read the buret again. The exact amount released is equal to the difference between your first and final buret reading. Practice measuring liquids by measuring 10 mL of water, using a graduated cylinder, a pipet, and a buret.

6. At the end of this part of the experiment, the equipment you store in the lab locker or drawer should be clean, dry, and arranged in an orderly fashion for the next lab experiment.
   CAUTION In many experiments you will have to dispose of a liquid chemical at the end of a lab. Always ask your teacher for the correct method of disposal. In many instances liquid chemicals can be washed down the sink's drain by diluting them with plenty of tap water. Toxic chemicals should be handled only by your teacher. All apparatus should be washed, rinsed, and dried.

7. Remember to wash your hands thoroughly with soap at the end of this part of the experiment.

MATERIALS

PART 6 FILTRATION

- 250 mL beakers, 2
- Bunsen burner with related equipment
- evaporating dish
- filter paper
- fine sand
- funnel
- glass stirring rod
- iron ring
- ring stand
- sodium chloride
- sparker
- wash bottle
- water
- wire gauze, ceramic-centered

EXPERIMENT B1 continued

PROCEDURE

1. Sometimes liquids contain particles of insoluble solids that are present either as impurities or as precipitates formed by the interaction of the chemicals used in the experiment. If the particles are denser than water, they soon sink to the bottom. Most of the clear, supernatant liquid above the solid may be poured off without disturbing the precipitate. This method of separation is known as decantation.

2. Fine particles or particles that settle slowly are often separated from a liquid by filtration. Support a funnel on a small ring on the ring stand as shown in Figure J. Use a beaker to collect the filtrate. Adjust the funnel so that the stem of the funnel just touches the inside wall of the beaker.

3. Fold a circular piece of filter paper along its diameter, and then fold it again to form a quadrant, as shown in Figure K. Separate the folds of the filter paper, with three thickness on one side and one on the other; then place the filter paper cone in the funnel.

FIGURE J

Filter paper folded in half

Filter paper folded in quarters

Filter paper ready for funnel

Filter paper in funnel

FIGURE K

EXPERIMENT B1 continued

The funnel should be wet before you insert the filter paper. Use your plastic wash bottle to wet the funnel and the filter paper. Press the edges of the filter paper firmly against the sides of the funnel so no air can get between the funnel and the filter paper while the liquid is being filtered. *EXCEPTION: A filter should not be wet with water when the liquid to be filtered does not mix with water. Why?*

Because the liquid does not mix with water, a filter paper wet with water will not allow the liquid to go through the filter paper.

4. Dissolve 2 or 3 g of salt in a beaker containing about 50 mL of water and stir into the solution an equal volume of fine sand. Filter out the sand by pouring the mixture into the filter, observing the following suggestions:
   a. The filter paper should not extend above the edge of the funnel. Use filter paper that leaves about 1 cm of the funnel exposed.
   b. Do not completely fill the funnel. It must never overflow.
   c. Try to establish a water column in the stem of the funnel to eliminate air bubbles, and then add the liquid quickly enough to keep the mixture level about 1 cm from the top of the filter paper.
   d. When a liquid is poured from a beaker, it may adhere to the glass and run down the outside wall. This may be avoided by holding a stirring rod against the lip of the beaker, as shown in Figure J on the previous page. The liquid will run down the rod and drop off into the funnel without running down the outside of the beaker. The sand is retained on the filter paper. What property of the sand enables it to be separated from the water by filtration?

   The sand is insoluble in water.

   What does the filtrate contain?

   The filtrate contains salt dissolved in water.

5. The salt can be recovered from the filtrate by pouring the filtrate into an evaporating dish and evaporating it over a low flame nearly to dryness. Figure L on the next page shows a correct setup for evaporation.
   CAUTION When using a Bunsen burner, confine loose clothing and long hair. Wear your safety goggles, lab apron, and gloves.

EXPERIMENT B1 continued

6. Remove the flame as soon as the liquid begins to spatter. Shut off the burner. What property of salt prevents it from being separated from the water by filtration?

   Salt is soluble in water.

7. All equipment should be clean, dry, and put away in an orderly fashion for the next lab experiment. Be sure that the valve on the gas jet is completely shut off. Make certain that the filter papers and sand are disposed of in the waste jars or containers and not down the sink. Remember to wash your hands thoroughly with soap at the end of each laboratory period.

FIGURE L

QUESTIONS

Answer the following questions in complete sentences.

1. **Organizing Ideas** As soon as you enter the lab, what safety equipment should you put on immediately?

   Safety goggles and lab apron should be put on.

2. **Organizing Ideas** Before doing an experiment, what should you read and discuss with your teacher?

   All safety precautions in the procedures should be read and discussed with the teacher.

3. **Organizing Ideas** Before you light a burner, what safety precautions should always be followed?

   Safety goggles and lab apron should be worn. Long hair should be tied back and loose clothing should be confined. One should know the locations of fire extinguishers, fire blankets, and sand buckets and the procedure for using them in case of a fire.

EXPERIMENT B1 continued

4. **Organizing Ideas** What immediate action should you take when the flame of your burner is burning inside the base of the barrel?

   Immediately turn off the gas at the gas valve.

   Allow the barrel to cool.

5. **Organizing Ideas** What type of flame is preferred for laboratory work and why?

   The nonluminous flame with the light blue inner cone is

   preferred because it burns hotter and cleaner than the

   yellow luminous flame.

6. **Analyzing Ideas** When inserting glass tubing, why is it important that you wear safety goggles and gloves or cover the tubing and stopper with protective pads of cloth?

   These precautions are necessary to prevent personal injuries

   such as cuts if the tubing should break. Eyes also need

   protection in case the tubing should break and scatter fragments.

7. **Analyzing Ideas** What do you think might be the common cause of fires in lab drawers or lockers?

   Putting away hot apparatus before it has had a chance to cool.

8. **Analyzing Ideas** Why are broken glassware, chemicals, matches, and other laboratory debris never discarded in a wastepaper basket?

   This is an important safety precaution against fires

   and personal injuries, such as cuts from broken glass or chemical

   poisoning, to anyone who empties the wastepaper basket.

EXPERIMENT B1 continued

9. **Organizing Ideas** List the safety precautions that should be observed when inserting, or removing glass tubing from a rubber stopper or rubber hose.

    1. The ends of the glass tubing must be fire polished.

    2. A lubricant should be used on the ends of the tubing.

    3. Always aim the glass tubing away from the palm of the hand that holds the stopper or hose.

    4. Put folded cloth pads between both hands and the glass tubing or wear leather gloves.

    5. Use a gentle twisting motion when inserting the tubing.

    6. Immediately remove the tubing at the end of an experiment.

    7. Follow the same precautions for removal as specified for insertion.

10. **Analyzing Ideas** Why should you never touch chemicals with your hands?

    Some chemicals readily pass through the skin into the bloodstream. Some chemicals are extremely corrosive.

11. **Organizing Ideas** What precaution can help prevent chemical contamination in reagent bottles?

    Never return unused chemicals to reagent bottles.

12. **Analyzing Ideas** Why are chemicals and hot objects never placed directly on the balance pan?

    They can permanently damage the surface of the balance pan and affect the accuracy of mass measurements.

ChemFile   EXPERIMENT B1   **15**

EXPERIMENT B1 continued

13. Organizing Ideas List three pieces of equipment used in the laboratory for measuring small quantities of liquids. What is the correct procedure for filling a buret with liquid?

Graduated cylinders, pipets, and burets are used to measure liquids. To fill a buret, first pour the liquid into a small beaker that is easy to handle. Then pour the liquid from the small beaker into the buret via a glass funnel.

14. Organizing Ideas What is the rule about the size of filter paper to be used with a funnel?

The filter paper should not extend above the edge of the funnel. Use filter paper that leaves about 1 cm of the funnel exposed above the edge of the filter cone.

15. Organizing Ideas How can a liquid be transferred from a beaker to a funnel without spattering and without running down the outside wall of the beaker?

A stirring rod may be held against the lip of the beaker. When you pour, the liquid will cling to the rod, run down it, and drop into the funnel.

16. Organizing Ideas Describe the condition of all lab equipment at the end of an experiment. What should be checked before you leave the lab?

All equipment that is stored in the lab locker or drawer should be completely cool, clean, dry, and arranged in an orderly fashion for the next lab experiment. Check to see that the valve on the gas jet is completely turned off.

EXPERIMENT B1 continued

17. Organizing Ideas What is the correct procedure for removing a solid reagent from its container in preparation for use in an experiment?

First, read the label on the reagent bottle before removing any of its contents. Use a clean spatula to remove the approximate quantity needed. Place the reagent on a piece of glazed paper or in a clean, premassed container, as directed by your teacher. Never touch the chemical with your hands and never pour the chemical from its original bottle into another container. Dispose of any unused chemical as directed by your teacher. Never return it to its original container.

18. Organizing Ideas What is the correct procedure for removing a liquid reagent from its container in preparation for use in an experiment?

First, read the label on the reagent bottle before removing any of its contents. When removing the glass stopper from a reagent bottle, never place the stopper on the lab table. Carefully pour the approximate amount of chemical needed into a small, clean, and dry beaker (50 mL) that is easy to handle. Then carefully transfer the liquid into the appropriate devise to measure the volume needed. When transferring to a buret, a funnel may be necessary to avoid spillage. The same cautions for using a solid reagent also apply to liquids.

19. Analyzing Ideas Why is it important to use low flame when evaporating water from a recovered filtrate?

A low flame will heat the evaporating dish and its contents gently. This will prevent the evaporating dish from cracking and will also keep the recovered solid from burning,

ChemFile   EXPERIMENT B1

EXPERIMENT B1 continued

GENERAL CONCLUSIONS

Safety Check

Identify the following safety symbols:

a. _eye safety_

b. _heating and fire safety_

c. _gas precaution_

d. _clothing protection_

e. _chemical safety_

f. _glassware safety_

g. _hand safety_

h. _explosion danger_

i. _radiation precaution_

j. _caustic substances_

Labeling

Practice labeling a chemical container or bottle by filling in the appropriate information missing on the label pictured on the following page. Use 6 M sodium hydroxide (NaOH) as the solution to be labeled. (Hint: 6 M sodium hydroxide is a caustic and corrosive solution and can be considered as potentially hazardous as 6 M HCl.

EXPERIMENT B1 continued

SAMPLE LABEL

- Working name of contents
- Chemical composition or formula

Sodium hydroxide

6 M NaOH

WARNING! CAUSTIC and CORROSIVE—CAUSES BURNS
CAUTION! Avoid contact with skin and eyes. Avoid breathing vapors

IN CASE OF CONTACT: Immediately flush skin or eyes with large amounts of water for at least 15 minutes: for eyes, get medical attention immediately!

Answers will vary.
Answers will vary.

- Statements of possible hazards and precautionary measures
- Date prepared or obtained
- Manufacturer (commercially obtained) or preparer (locally made)
- Instructions in case of contact or exposure

True or False

Read the following statements and indicate whether they are true or false. Place your answer in the space next to the statement.

Answer	Statement
True	1. Never work alone in the laboratory.
True	2. Never lay the stopper of a reagent bottle on the lab table.
False	3. At the end of an experiment, in order to save the school's money, save all excess chemicals and pour them into their stock bottles.
False	4. The quickest and safest way to heat a material in a test tube is by concentrating the flame on the bottom of the test tube.
True	5. Use care in selecting glassware for high-temperature heating. Glassware should be Pyrex or a similar heat-treated type.
True	6. A mortar and pestle should be used for grinding only one substance at a time.
False	7. Safety goggles protect your eyes from particles and chemical injuries. It is completely safe to wear contact lenses under them while performing experiments.
True	8. Never use the wastepaper basket for disposal of chemicals.
False	9. First aid kits may be used by anyone to give emergency treatment after an accident.
True	10. Eyewash and facewash fountains and safety showers should be checked daily for proper operation.

EXPERIMENT B1 continued

Chemical Apparatus

Identify each piece of apparatus. Place your answers in the spaces provided.

a.
b.
c.
d.
e.
f.
g.
h.
i.
j.
k.
l.

a. beaker

b. graduated cylinder

c. spatula

d. pipestem triangle

e. funnel

f. evaporating dish

g. Erlenmeyer flask

h. Bunsen burner

i. test tube

j. ceramic-centered wire gauze

k. crucible and cover

l. mortar and pestle

EXPERIMENT B2

Accuracy and Precision: Calibrating a Pipet

OBJECTIVES

Recommended time:
60 min

Required Precautions
- Read all safety cautions and discuss them with your students.

- **Use** experimental measurements in calculations.
- **Organize** data by compiling it in tables.
- **Compute** an average value from class data and use the value to calculate absolute deviation and average deviation.
- **Recognize** the importance of accuracy and precision in scientific measurements.
- **Relate** the reliability of experimental data to absolute deviation, average deviation, uncertainty, and percent error.

INTRODUCTION

- Safety goggles and a lab apron must be worn at all times.
- Tie back loose hair and long clothing when working in the lab.

Procedural Tips
- Students who have done Experiment 1-1 will already know how to read volume gradations on a graduated cylinder and use the balance. They may only need help in reading more precisely.

In this experiment, you will determine the volume of a liquid in two different ways and compare the results. You will also calculate the density of a metal using your measurements of its mass and volume.

$$D = \frac{m}{V}$$

You will compare your result with the accepted value found in a handbook. The error and percent error in each part of the experiment will be calculated.

The *experimental error* is calculated by subtracting the accepted value from the observed or experimental value. The *percent error* is calculated according to the following equation.

$$\text{Percent error} = \frac{\text{Observed value} - \text{Accepted value}}{\text{Accepted value}} \times 100$$

The sign of the experimental error and the percent error may be either positive (the experimental result is too high) or negative (the experimental result is too low).

You will average the values for the density of a metal obtained by the entire class to determine the average value. Using this value you will calculate the *average deviation* of these data. The average deviation will be expressed as the *uncertainty* of the measurements.

SAFETY

Always wear safety goggles and a lab apron to protect your eyes and clothing. If you get a chemical in your eyes, immediately flush the chemical out at the eyewash station while calling to your teacher. Know the location of the emergency lab shower and eyewash station and the procedure for using them.

EXPERIMENT B2 continued

Do not touch any chemicals. If you get a chemical on your skin or clothing, wash the chemical off at the sink while calling to your teacher. Make sure you carefully read the labels and follow the precautions on all containers of chemicals that you use. If there are no precautions stated on the label, ask your teacher what precautions to follow. Never return leftover chemicals to their original containers; take only small amounts to avoid wasting supplies.

Never put broken glass into a regular waste container. Broken glass should be disposed of separately according to your teacher's instructions.

MATERIALS

- 15 cm plastic ruler
- 25 mL graduated cylinder
- 100 mL beaker
- 100 mL graduated cylinder
- balance
- metal shot (aluminum, copper, lead)
- thermometer, nonmercury, 0–100°C

PROCEDURE

- Students may have difficulty understanding the difference between accuracy and precision. You might use the analogy of the dart board. If darts puncture the board over a large circular area surrounding the bull's eye, the shots may be considered to be accurate because an average of the positions of all the shots would come close to the position of the bull's eye. However, the shots would not be precise because they are not close together. On the other hand, a distribution of shots such that they are clustered in a small area below the bull's eye might be considered precise because they are close together, but not accurate because their positions do not average to that of the bull's eye.

RECORDING YOUR OBSERVATIONS

After completing each part of the experiment, record your observations in the appropriate data table. Recording data anywhere else increases the probability of recording an inaccurate value.

Part 1

1. Examine the centimeter scale of the plastic ruler. What are the smallest divisions?

   The smallest divisions on the ruler are millimeters.

   To what fraction of a centimeter can you make measurements with such a ruler?

   You should be able to measure to ± 0.05 cm.

2. Using the ruler, measure the inside diameter of the 100 mL graduated cylinder. Similarly, measure the inside height of the cylinder to the 50 mL mark. Record these measurements in Data Table 1.

   Data Table 1

Inside diameter of graduated cylinder	2.50 cm
Inside height of graduated cylinder	9.31 cm

EXPERIMENT B2 continued

- Impress upon students that an estimate of uncertainty is a necessary part of every experimental measurement, that the uncertainty determines the number of significant digits the measurement can have, and that uncertainty increases as measurements are added and subtracted, multiplied and divided.

Disposal

The metal shot can be dried and reused in each class.

Part 2

3. Examine the gram scale of the balance. What are the smallest divisions?

The smallest divisions are centigrams.

To what fraction of a gram can you make measurements with a centigram balance?

You should be able to measure to ± 0.01 g.

4. Examine the graduations on a 25 mL graduated cylinder and determine the smallest fraction of a milliliter to which you could make a measurement. Does this match the uncertainty of a measurement made with a 100 mL graduated cylinder?

Most 25 mL cylinders are subdivided by 0.5-mL

gradations, whereas 100 mL cylinders are subdivided

by 1-mL gradations. The uncertainties of measurements

using the two cylinders do not match (± 0.25 mL for the

25 mL cylinder and ± 0.5 mL for the 100 mL cylinder).

5. Using the balance, determine the mass of the dry 25 mL cylinder. Record the mass in Data Table 2.

6. Fill the beaker half full of water and determine its temperature to the nearest degree. Look up the density of water for this temperature, and record in the data table both the temperature and water density.

7. Fill your graduated cylinder with water to a level between 10 and 25 mL; accurately read and record the volume. Determine the mass of the water plus the cylinder. Then, record this value in Data Table 2. Save the water in the graduated cylinder for use in Part 3.

Data Table 2

Mass of empty graduated cylinder	42.39	g
Temperature of water	25.0	°C
Density of water	0.997	g/cm³
Volume of water	13.80	mL
Mass of graduated cylinder + water	55.98	g

ChemFile

EXPERIMENT B2

EXPERIMENT B2 continued

Part 3

8. Add a sufficient quantity of the assigned metal shot (aluminum, copper, or lead) to the cylinder containing the water (saved from Part 2) to increase the volume by at least 5 mL. Determine the volume and then the mass of the shot, water, and cylinder. Record your measurements in Data Table 3.

Data Table 3

Volume of water (from Part 2)	13.80	mL
Mass of water + graduated cylinder (from Step 2)	55.98	g
Volume of metal and water	20.25	mL
Mass of metal + water + graduated cylinder	73.98	g

Cleanup and Disposal

9. Clean up all apparatus and your lab station. Return equipment to its proper place. Dispose of chemicals and solutions in the containers designated by your teacher. Do not pour any chemicals down the drain or in the trash unless your teacher directs you to do so. Wash your hands thoroughly with soap before you leave the lab and after all work is finished.

CALCULATIONS

Show all your calculations. Place your answers in the appropriate calculations table.

Part 1

1. **Organizing Data** Calculate the volume of the cylinder to the 50.0-mL graduation ($V = 3.14 \times r^2 \times h$).

$$V = 3.14\, r^2 h = 3.14 \times \left(\frac{2.50 \text{ cm}}{2}\right)^2 \times 9.31 \text{ cm} = 45.7 \text{ cm}^3$$

2. **Inferring Conclusions** Assume the accepted value is 50.0 cm³. Calculate the error and percent error.

Error: 45.7 cm³ − 50.0 cm³ = 4.3 cm³

Percent error: $\dfrac{4.3 \text{ cm}^3}{50.0 \text{ cm}^3} \times 100 = 8.6\%$

Part 2

3. **Organizing Data** Calculate the mass of water as measured by the balance.

(Mass of cylinder + water) − mass of cylinder = mass of water

55.98 g − 42.39 g = 13.59 g of water

EXPERIMENT B2 continued

4. **Organizing Data** Calculate the mass of the water from its measured volume and its density ($m = D \times V$).

    $m = D \times V = 0.997 \text{ g/cm}^3 \times 13.80 \text{ cm}^3 = 13.8 \text{ g}$

5. **Inferring Conclusions** Using the mass of water determined by the use of the balance as the *accepted value,* calculate the error and percent error in the mass determined using the volume and density.

    Error: $13.80 \text{ g} - 13.59 \text{ g} = 0.21 \text{ g}$

    Percent error: $\dfrac{0.21 \text{ g}}{13.59 \text{ g}} \times 100 = 1.5\%$

Part 3

6. **Organizing Data** Determine the volume of the metal using your measurement of the volume of water displaced by the metal.

    Volume of metal = $20.25 \text{ mL} - 13.80 \text{ mL} = 6.45 \text{ mL} = 6.45 \text{ cm}^3$

7. **Organizing Data** Using your measurements in Data Table 3, determine the mass of the metal.

    Mass of metal = $73.98 \text{ g} - 55.98 \text{ g} = 18.00 \text{ g}$

8. **Organizing Data** Calculate the density of the metal.

    Density of metal = $\dfrac{18.00 \text{ g}}{6.45 \text{ cm}^3} = 2.79 \text{ g/cm}^3$

9. **Inferring Conclusions** In a handbook or the appendix of your textbook, look up the specific gravity of the metal you used. (In the metric system, the density of liquids and solids is equal to the specific gravity.) Calculate the error and percent error for the density of the metal shot you determined in item **8.**

    Experimental value − accepted value = error

    Error: $2.79 \text{ g/cm}^3 - 2.7 \text{ g/cm}^3 = 0.1 \text{ g/cm}^3$

    Percent error: $\dfrac{0.1 \text{ g/cm}^3}{2.7 \text{ g/cm}^3} \times 100 = 4\%$

EXPERIMENT B2 continued

Part 4

10. Organizing Conclusions Record in the table below five values obtained by you and your classmates for the density of the same metal.

Group Number	Density (g/cm³)
1	2.79
2	3.00
3	2.95
4	2.79
5	2.81

11. Evaluating Conclusions Calculate the average density (M) of the five results.

$$M = \frac{2.79 + 3.00 + 2.95 + 2.79 + 2.81}{5} = 2.87 \text{ g/cm}^3$$

QUESTIONS

1. Evaluating Methods What value of a measurement must be known if the accuracy of an experimental measurement is to be determined?

The accepted value of a measurement must be known.

2. Evaluating Methods What are the possible sources of experimental errors in this experiment?

Sources of experimental errors include inaccurate observations and imperfections in apparatus design.

GENERAL CONCLUSIONS

1. Evaluating Conclusions Sarah and Jamal determined the density of a liquid three times. The values they obtained were 2.84 g/cm³, 2.85 g/cm³, and 2.80 g/cm³. The accepted value is known to be 2.40 g/cm³.

a. Are the values that Sarah and Jamal determined precise? Explain.

Yes, the measurements are in close agreement

EXPERIMENT B2 continued

    b. Are the values accurate? Explain.

    No, they are not close enough to the real value.

    c. Calculate the percent error and the uncertainty for each measurement.

Uncertainty: Average Density 2.83 g/cm³

Experiment	Experimental density (g/cm³)	Average density (g/cm³)	Absolute deviation g/cm³
1	2.84	2.83	0.1
2	2.85	2.83	0.2
3	2.80	2.83	0.3

Percent error

1. $\dfrac{2.84 \text{ g/cm}^3 - 2.40 \text{ g/cm}^3}{2.40 \text{ g/cm}^3} \times 100 = 18.3\%$

2. $\dfrac{2.85 \text{ g/cm}^3 - 2.40 \text{ g/cm}^3}{2.40 \text{ g/cm}^3} \times 100 = 18.8\%$

3. $\dfrac{2.80 \text{ g/cm}^3 - 2.40 \text{ g/cm}^3}{2.40 \text{ g/cm}^3} \times 100 = 16.7\%$

Name _____

Date _____ Class _____

HOLT ChemFile
LAB PROGRAM

EXPERIMENT B3

Relative Solubility of Transition Elements

OBJECTIVES

Recommended time: 50 min

- **Observe** the reactions of some transition-element ions with five anions.
- **Deduce** the solubility of transition-element compounds in water and in acid.
- **Compare** the solubilities of elements in the same column and in the same row of the periodic table.

INTRODUCTION

Solution/Material Preparation

1. To prepare 100 mL of 0.1 M copper nitrate, dissolve 2.4 g of $Cu(NO_3)_2 \cdot 3H_2O$ in enough distilled water to make 100 mL of solution.

2. To prepare 100 mL 0.1 M iron(III) nitrate, dissolve 3.5 g of $Fe(NO_3)_3 \cdot 6H_2O$ in enough distilled water to make 100 mL of solution.

The transition elements are found in Periods 4, 5, and 6 between Groups 2 and 13 of the periodic table. As the atomic number increases across a row in this section of the table, no new valence electrons are added to the highest energy level. Instead, additional electrons are added to inner energy levels. As a result, the properties of the transition elements are similar across a row as well as down a column. In many instances, the properties of the elements in a given row are more alike than the properties of elements in a given column. This is not the case for nontransition elements.

In this experiment, you will evaluate the solubility of compounds of iron, copper, zinc, and mercury in water and in acid. Iron, copper, and zinc are in the same row of the periodic table; zinc and mercury are in the same column. You will mix soluble ionic compounds of these transition elements with five different reagents and observe whether an insoluble compound (precipitate) forms. For those mixtures that form precipitates, you will investigate how the precipitate behaves in acid. After tabulating your findings, you will decide whether each new compound is soluble or insoluble in water and to what extent it dissolves in acid. You can then look for trends related to the positions of the elements in the periodic table.

SAFETY

3. To prepare 100 mL of 0.1 M mercury(I) nitrate, dissolve 5.6 g of $Hg_2(NO_3)_2 \cdot 2H_2O$ in enough distilled water to make 100 mL of solution.

4. To prepare 25 mL of 0.1 M zinc nitrate, dissolve 0.6 g of $Zn(NO_3)_2 \cdot 3H_2O$ in enough distilled water to make 25 mL of solution.

For this experiment, wear safety goggles, gloves, and a lab apron to protect your eyes, hands, and clothing. If you get a chemical in your eyes, immediately flush the chemical out at the eyewash station while calling to your teacher. Know the location of the emergency lab shower and eyewash station and the procedure for using them.

Do not touch any chemicals. If you get a chemical on your skin or clothing, wash the chemical off at the sink while calling to your teacher. Make sure you carefully read the labels and follow the precautions on all containers of chemicals that you use. If there are no precautions stated on the label, ask your teacher what precautions you should follow. Do not taste any chemicals or items used in the laboratory. Never return leftovers to their original containers; take only small amounts to avoid wasting supplies.

ChemFile

EXPERIMENT B3

HRW material copyrighted under notice appearing earlier in this work.

EXPERIMENT B3 continued

Call your teacher in the event of a spill. Spills should be cleaned up promptly, according to your teacher's directions.

MATERIALS

5. To prepare 100 mL of 1.0 M nitric acid, observe the required safety precautions. Slowly and with stirring, add 6.4 mL of concentrated HNO₃ to enough distilled water to make 100 mL of solution.

6. To prepare 100 mL of 0.1 M sodium chloride, dissolve 0.6 g of NaCl to enough distilled water to make 100 mL of solution.

- 24-well microplate or plastic sheet
- reagents
  0.1 M $K_4Fe(CN)_6$
  0.1 M KSCN
  0.1 M NaCl
  0.1 M Na_2SO_4
  0.2 M $(NH_4)_2C_2O_4$
  1.0 M HNO_3
- transition elements
  0.1 M $Cu(NO_3)_2$
  0.1 M $Fe(NO_3)_3$
  0.1 M $Hg_2(NO_3)_2$
  0.1 M $Zn(NO_3)_2$
- thin-stemmed pipets or dropper bottles for solutions, 10

PROCEDURE

7. To prepare 100 mL of 0.2 M ammonium oxalate, dissolve 2.8 g of $(NH_4)_2C_2O_4 \cdot H_2O$ in enough distilled water to make 100 mL of solution.

8. To prepare 100 mL of 0.1 M sodium sulfate, dissolve 1.4 g of Na_2SO_4 in enough distilled water to make 100 mL of solution.

9. To prepare 100 mL of potassium thiocyanate, dissolve 1.0 g of KSCN in enough distilled water to make 100 mL of solution.

10. To prepare 100 mL of 0.1 M potassium hexacyanoferrate(II), dissolve 4.2 g of $K_4Fe(CN)_6 \cdot 3H_2O$ in enough distilled water to make 100 mL of solution.

11. A 10 in. square piece of plexiglass is an ideal surface for the drops.

1. Construct a full-page grid similar to the Solubility Chart that precedes the Data Table. Place a microplate or plastic sheet over the grid.

2. Place 2 drops of each solution containing a transition-element ion (positive ion) on the plastic sheet in the appropriate box. There should be five boxes in a row for each transition element. Add 2 drops of each of the reagents (negative ions) to each of the transition-element ions. **Do not allow the dropper to touch the transition-element solutions.**

3. Immediately after adding the drops of reagent ion to the transition ion, record your observations in the section of the data table labeled *"0 min."* If no reaction occurs, write *NR*. If a precipitate forms, record its color.

4. Wait 5 min, and then record any changes in the mixtures in the section of the data table labeled *"5 min."* If no change occurs, write *same*.
   CAUTION Nitric acid is corrosive and caustic. Avoid contact with eyes and skin. If any should spill on you, immediately flush the area with water and notify your teacher.

5. To each mixture in which a precipitate formed, add 1 to 4 drops of 1.0 M nitric acid. Write *clear* in the *"Acid"* section of the data table if the precipitate disappeared. Write *PC* if the precipitate partially disappeared and *same* if there was no change.

Cleanup and Disposal

6. Clean all apparatus and your lab station. Return equipment to its proper place. Dispose of chemicals and solutions in the containers designated by your teacher. Do not pour any chemicals down the drain or into the trash unless your teacher directs you to do so. Wash your hands thoroughly before you leave the lab and after all work is finished.

EXPERIMENT B3 continued

Required Precautions

- Safety goggles and a lab apron must be worn at all times.
- Read all safety precautions, and discuss them with your students.
- In case of a spill, use a dampened cloth or paper towel (or more than one towel if necessary) to mop up the spill. Then rinse the towel in running water at the sink, wring it out until it is only damp, and put it in the trash. In the event of a spill of an acid, dilute the spill with water and then proceed as described.
- Wear safety goggles, a face shield, impermeable gloves, and a lab apron when you prepare the HNO_3. Work in a hood known to be in operating condition, with another person present nearby to call for help in case of an emergency. Be sure you are within a 30 s walk of a safety shower and eyewash station known to be in operating condition.

Pre-lab Discussion

Make sure students understand that when a precipitate forms it happens because a new compound has formed from the mixing of the two solutions. Because the new compound has low solubility, it cannot remain in solution. Caution students not to contaminate the solutions by touching the drops with the pipet.

QUESTIONS

Data Table

Transition ions		Cl$^-$	C$_2$O$_4$$^{2-}$	SO$_4$$^{2-}$	SCN$^-$	Fe(CN)$_6$$^{4-}$
Cu^{2+}	0 min	NR	light aqua ppt	NR	NR	red-orange ppt
	5 min	same	same	same	same	same
	acid		clear			same
Fe^{3+}	0 min	NR	NR	NR	NR	deep blue ppt
	5 min	same	same	same	same	same
	acid					
Hg$_2$$^{2+}$	0 min	white ppt	white ppt	yellow-brown ppt	gray ppt	yellow-green ppt
	5 min	same	same	same	same	blue-green ppt
	acid	PC	PC	clear	PC	bright blue ppt
Zn^{2+}	0 min	NR	clear	NR	NR	gel
	5 min	same	light white ppt	same	same	same
	acid		clear			same

Reagent ions

Solubility Chart

Transition Ions	Cl$^-$	C$_2$O$_4$$^{2-}$	SO$_4$$^{2-}$	SCN$^-$	Fe(CN)$_6$$^{4-}$
Cu^{2+}	S	SA	S	S	I
Fe^{3+}	S	S	S	S	I
Hg$_2$$^{2+}$	PSA	PSA	SA	PSA	I
Zn^{2+}	S	PS	S	S	I

Reagents

1. Organizing Data and Analyzing Results Study your data table carefully. Then complete the solubility chart that follows the data table. For each combination of positive transition ions and negative reagent ions, write *S* if the mixture is soluble in water (no reaction) and *PS* if it is partially soluble in water. Write *SA* if the mixture is soluble in acid but not in water and *PSA* if it is partially soluble in acid but not in water. Write *I* if the mixture is insoluble in water and acid. Assume a mixture is partially soluble in water if it was clear at 0 min but became cloudy after 5 min. Assume a mixture is partially soluble in acid if the precipitate partially dissolved after the addition of HNO_3.

EXPERIMENT B3 continued

Disposal

- Combine all solutions containing Fe³⁺. Add an excess of potassium ferrocyanide to precipitate all Fe³⁺. Filter and throw the precipitate into the trash. Adjust the pH of the filtrate to between 5 and 9 with NaOH and pour the mixture down the drain.
- Combine all Zn²⁺ solutions. Add 1.0 M NaOH, slowly, with stirring, to precipitate all of the zinc as a hydroxide.
- Combine all solutions containing mercury in a separate container. Use your local guidelines for disposal.

GENERAL CONCLUSIONS

- Combine all solutions containing Cu²⁺ in a separate container. Add water and allow the solution to evaporate. Recycle as a source of copper. Do not dispose of this waste in a landfill, incinerator, or down the drain.

2. **Analyzing Results** Which transition ion(s) form(s) the greatest number of soluble compounds with the five negative ions used in this experiment?

   Iron forms four soluble compounds; copper and zinc each form three.

3. **Analyzing Results** Which transition ion(s) form(s) the smallest number of soluble compounds with the five negative ions used in this experiment?

   mercury

4. **Analyzing Results** Zinc and mercury are in the same vertical column of the periodic table. Iron, copper, and zinc are in the same horizontal row. According to your solubility chart, are the solubilities more similar across the row, down the column, or neither?

   Overall, solubilities are more similar across the row.

1. **Analyzing Information** Use your data table and solubility chart to identify each of the following unknown ions.
   a. Forms a white precipitate with Cl⁻

   Hg_2^{2+}

   b. Forms a precipitate with all five reagent ions

   Hg_2^{2+}

   c. Two reagent ions that can distinguish Hg_2^{2+} from the other three transition-element ions

   Cl⁻ and SCN⁻

   d. The reagent ion that forms the most insoluble transition-metal compounds

   $Fe(CN)_6^{4-}$

2. **Predicting Outcomes** Notice the position of cadmium in the periodic table. On the basis of your data on zinc and mercury, predict the overall solubility of cadmium.

   Students may answer that the overall solubility of cadmium is intermediate between that of zinc and mercury. Cadmium chloride, sulfate, and thiocyanate are soluble in water. Cadmium forms a gel with hexacyanoferrate(II), which becomes a white precipitate in acid solution. Cadmium oxalate is insoluble in water but soluble in acid.

EXPERIMENT B4

Periodicity of Properties of Oxides

OBJECTIVES

Recommended time:
50 min

- **Determine** the solubility in water of several oxides.
- **Classify** these oxides as acidic or basic.
- **Compare** the solubility of two oxides in strong acid and in strong base.
- **Classify** these oxides as acidic, basic, or amphoteric.
- **Relate** experimental results to the positions of the elements in the Periodic Table and generalize trends.

INTRODUCTION

Materials (for quantities not in student's list)

- 6 M HCl, 5 mL
- 6 M NaOH, 5 mL
- MgO, 0.25 g
- SnO$_2$, 0.25 g
- SrCO$_3$, 0.5 g
- ZnO, 0.25 g
- carbonated beverage, 5 mL
- sulfur, 1 g
- universal indicator, 2 mL

Some oxides produce acidic solutions when they dissolve in water. These oxides are classified as acidic oxides (acid anhydrides), and they are the primary cause of acid rain. Other oxides produce basic solutions when they dissolve in water. These oxides are known as basic oxides (basic anhydrides). In general, nonmetal oxides produce acidic solutions and metallic oxides produce basic solutions.

To classify oxides as acidic or basic, the solubility of the oxide must be tested in both a strong base and a strong acid. If the oxide is more soluble in a strong base such as sodium hydroxide, the oxide is classified as an acidic oxide. If the oxide is more soluble in a strong acid such as hydrochloric acid, the oxide is classified as a basic oxide. Those oxides that dissolve in both an acid and a base are called amphoteric oxides.

In this experiment you will investigate the properties of a number of oxides. From your observations and the relative positions of the oxides in the periodic table, you can generalize group and period trends in these properties.

SAFETY

Solution/Material Preparation

1. To prepare 100 mL of 6 M hydrochloric acid, observe the required safety precautions. Slowly and with stirring, add 52 mL of concentrated HCl to enough distilled water to make 100 mL of solution.

Always wear safety goggles and a lab apron to protect your eyes and clothing. If you get a chemical in your eyes, immediately flush the chemical out at the eyewash station while calling to your teacher. Know the location of the emergency lab shower and eyewash station and the procedure for using them.

Do not touch any chemicals. If you get a chemical on your skin or clothing, wash the chemical off at the sink while calling to your teacher. Make sure you carefully read the labels and follow the precautions on all containers of chemicals that you use. If there are no precautions stated on the label, ask your teacher what precautions you should follow. Do not taste any chemicals or items used in the laboratory. Never return leftovers to their original containers; take only small amounts to avoid wasting supplies.

EXPERIMENT B4 continued

2. To prepare 100 mL of 6 M sodium hydroxide, observe the required safety precautions. Slowly and with stirring, dissolve 24 g of solid NaOH in enough distilled water to make 100 mL of solution.

When using a Bunsen burner, confine long hair and loose clothing. Do not heat glassware that is broken, chipped, or cracked. Use tongs or a hot mitt to handle heated glassware and other equipment; heated glassware does not always look hot. If your clothing catches fire, WALK to the emergency lab shower and use it to put out the fire.

Call your teacher in the event of a spill. Spills should be cleaned up promptly, according to your teacher's directions.

Never put broken glass in a regular waste container. Broken glass should be disposed of separately. Put the glass and ceramic pieces in the container designated by your teacher.

MATERIALS

Required Precautions

- Safety goggles and a lab apron must be worn at all times.
- Read all safety precautions, and discuss them with your students.
- In case of a spill, use a dampened cloth or paper towel (or more than one towel if necessary) to mop up the spill. Then rinse the towel in running water at the sink, wring it out until it is only damp, and put it in the trash. In the event of a spill of an acid, dilute the spill with water and then proceed as described.

- 6 M HCl
- 6 M NaOH
- 24-well microplate
- 50 mL beaker
- 250 mL gas collecting bottle
- Bunsen burner and heating set-up
- carbonated beverage
- crucible
- deflagrating spoon
- hot plate
- index card, 3 in. × 5 in.
- MgO
- rubber stopper for bottle
- SnO_2
- $SrCO_3$
- stirring rod
- sulfur
- thermometer, nonmercury
- thin-stemmed pipet
- thin-stemmed pipet with two right- angle bends
- universal indicator
- ZnO

PROCEDURE

1. Place a small sample (about the size of an apple seed) of magnesium oxide, MgO, into a well of the microplate. Using a thin-stemmed pipet, add deionized or distilled water until the well is half full. Stir the solution with a stirring rod. Add 3 drops of universal indicator solution. Record the result in Data Table 1.

2. Heat 0.5 g of strontium carbonate, $SrCO_3$, in a crucible until all the CO_2 has been driven off. Transfer some of the residue (SrO) to a well of the microplate, and repeat step **1**.

3. On a hot plate, heat 25 mL of water in a 50 mL beaker.

4. Using a thin-stemmed pipet, add deionized or distilled water to one well of the microplate until half full, and add 3 drops of universal indicator solution.

EXPERIMENT B4 continued

- Wear safety goggles, a face shield, impermeable gloves, and a lab apron when you prepare the HCl and NaOH. When preparing the HCl, work in a hood known to be in operating condition, with another person present nearby to call for help in case of an emergency. Be sure you are within a 30 s walk of a safety shower and eyewash station known to be in operating condition.

5. Obtain a thin-stemmed pipet with two right-angle bends. Fill this pipet half full with a carbonated beverage and make sure that no beverage remains in the stem. Rinse off the outside of the pipet with deionized water.

6. When the water in the beaker is between 70°C and 80°C, remove the beaker from the hot plate. Place the open end of the thin-stemmed pipet into the universal indicator solution in the well and set the bulb into the beaker of hot water, as shown in Figure A. Allow the CO_2 to bubble for approximately 30 s and record the results in Data Table 1.

FIGURE A

- Steps **10** and **11** are a teacher demonstration. Students should not perform these steps. Wear safety goggles, face shield, lab apron and gloves while performing these steps. Students should wear safety goggles and lab aprons, and stand behind a safety shield while observing. Be sure you are within a 30 s walk of a safety shower and eyewash station known to be in operating condition.

Techniques to Demonstrate

Show students how to set up the thin-stemmed pipet (with the two right-angle bends) for collecting and dissolving the CO_2 from the carbonated beverage. Caution students to keep the production and dissolving of SO_2 well inside the hood because the gas is unpleasant to breathe. Show them the set-up for burning the sulfur.

7. Fill a 50 mL beaker half full with tap water.
   CAUTION Because SO_2 is a very irritating gas, it should be prepared in a well-ventilated hood.

8. While working under a hood, pour 10 mL of deionized water into a gas-collecting bottle and add three drops of universal indicator solution. Fill a deflagrating spoon half full with sulfur and insert the handle through the center of an index card. Use a Bunsen burner to heat the spoon until the sulfur ignites. Quickly move the spoon into the center of the gas-collecting bottle, as shown in Figure B, and slide the index card down until it covers the opening of the bottle.

9. After the sulfur has burned for 30 s, remove the spoon and set it inside the 50 mL beaker of water to extinguish the sulfur. Stopper the bottle and shake to dissolve the SO_2. Use this solution to fill a well of the microplate half full. Record your results in Data Table 1.

FIGURE B

ChemFile   EXPERIMENT B4   **35**

EXPERIMENT B4 continued

Pre-lab Discussion

By recognizing the general rule that acidic oxides are formed by nonmetals, and basic oxides by metals, students can organize their results and fit them into an overall understanding of the periodic table. Encourage students to refer to the periodic table before they begin the lab work and in answering the questions.

Disposal

Combine all solids and put them into the trash. Combine all liquids, adjust the pH to between 5 and 9, and pour down the drain.

Your teacher will perform steps **10** and **11**.

CAUTION The teacher will wear safety goggles, a face shield, a lab apron, and gloves while performing this demonstration. Students should wear safety goggles and lab apron and stand behind a safety shield while observing. A safety shower and eyewash station, known to be in operating condition, should be within a 30 s walk of the demonstration.

10. To a small test tube containing ZnO, your teacher will add 20 drops of 6 M NaOH while stirring. To another test tube containing ZnO, your teacher will add 20 drops of 6 M HCl while stirring. If the oxide dissolves in a strong base, it is an acidic oxide; if it dissolves in a strong acid, it is a basic oxide. If the oxide dissolves in both a strong acid and a strong base, it is amphoteric. Record the results in Data Table 2.

11. Your teacher will repeat step **10** using SnO_2 instead of ZnO. Record the results.

Cleanup and Disposal

12. Clean all apparatus and your lab station. Return equipment to its proper place. Dispose of chemicals and solutions in the containers designated by your teacher. Do not pour any chemicals down the drain or into the trash unless your teacher directs you to do so. Wash your hands thoroughly before you leave the lab and after all work is finished.

Data Table 1

Oxide used	Universal indicator color	pH	Nature of oxide (acidic, basic, or amphoteric)
MgO	violet	11	basic
SrO	violet	11	basic
CO_2	orange red	3	acidic
SO_2	pink	1	acidic

Data Table 2

Oxide used	Solubility in 6 M HCl	Solubility in 6 M NaOH	Nature of oxide (acidic, basic, or amphoteric)
ZnO	soluble	soluble	amphoteric
SnO_2	soluble	soluble	amphoteric

QUESTIONS

1. **Organizing Results** For the oxides that you or your teacher tested, make a horizontal list of the group numbers of the elements forming the oxides, starting with Group 2. Write the formulas of the tested compounds under

EXPERIMENT B4 continued

their group numbers. Under each formula, write whether you found the oxide to be acidic, basic, or amphoteric.

2	12	14	16
MgO	ZnO	CO_2	SO_2
basic	amphoteric	acidic	acidic
SrO	SnO_2		
basic	amphoteric		

2. **Inferring Conclusions** What is the *general* trend in the properties of the oxides from left to right across the periodic table?

   Basic oxides on the left change to amphoteric, and then to acidic on the right.

3. **Analyzing Results** For the two elements you tested in Group 14, what is the trend in properties within the group?

   The change is from acidic at the top of the group (carbon) to amphoteric at the bottom (tin).

4. **Applying Ideas** Explain your answer to item 3 by classifying the two elements as metals or nonmetals.

   Carbon is a nonmetal; nonmetals form acidic oxides. Tin is a metal located in the table close to the boundary separating metals from nonmetals; it shows the properties of both.

EXPERIMENT B4 continued

5. Predicting Outcomes From what you have learned in this experiment, predict whether NO_2 is an acidic or a basic oxide. Explain your answer.

Nitrogen is in Group 15 between CO_2 and SO_2, both of which are acidic. Therefore, NO_2 is acidic.

6. Predicting Outcomes From what you have learned in this experiment, predict whether SnO or PbO is the stronger basic anhydride. Explain your answer.

PbO; acid strength of the oxides decreases from the top of the column to the bottom.

GENERAL CONCLUSIONS

1. Applying Conclusions Two test tubes containing insoluble oxides need to be cleaned. One contains a basic oxide, CaO, and the other contains an amphoteric oxide, As_4O_{10}. Suggest a way of cleaning these test tubes.

Because basic oxides dissolve in strong acids, add 6 M HCl to the CaO. Because amphoteric oxides dissolve in either a strong acid or base, add either 6 M NaOH or 6 M HCl to the As_4O_{10}.

2. Applying Conclusions In regions where soils tend to be too acidic for some plants, lime, CaO, is often added to the lawns and gardens. Explain how this can help.

When water dissolves the lime, a base is produced that neutralizes acids in the soil.

EXPERIMENT B5

Reactivity of Halide Ions

OBJECTIVES

Recommended time:
50 min

- **Observe** the reactions of the halide ions with different reagents.
- **Analyze** data to determine characteristic reactions of each halide ion.
- **Infer** the identity of unknown solutions.

INTRODUCTION

Materials (for quantities not in student's list)

- 0.1 M AgNO$_3$, 1 mL
- 0.5 M Ca(NO$_3$)$_2$, 1 mL
- 0.2 M KBr, 1 mL
- 0.2 M KI, 1 mL
- 0.1 M NaCl, 1 mL
- 0.1 M NaF, 1 mL
- 0.2 M Na$_2$S$_2$O$_3$, 1 mL
- 4 M NH$_3$(*aq*), 15 mL
- 5% NaOCl (commercial bleach), 1 mL
- 3% Starch solution, 1 mL

The four halide salts used in this experiment are found in your body. Although sodium fluoride is poisonous, trace amounts seem to be beneficial to humans in the prevention of tooth decay. Sodium chloride is added to most of our food to increase flavor while masking sourness and bitterness. Sodium chloride is essential for many life processes, but excessive intake appears to be linked to high blood pressure. Sodium bromide is distributed throughout body tissues, and in the past it has been used as a sedative. Sodium iodide is necessary for the proper operation of the thyroid gland, which controls cell growth. The concentration of sodium iodide is almost 20 times greater in the thyroid than in blood. The need for this halide salt is the reason that about 10 ppm of NaI is added to packages of table salt labeled "iodized."

The principal oxidation state of the halogens is −1. However, all halogens except fluorine may exist in other oxidation states. The specific tests you will develop in this experiment involve the production of recognizable precipitates and complex ions. You will use your observations to determine the halide ion present in an unknown solution.

SAFETY

Solution/Material Preparation

1. To prepare 50 mL of 0.1 M AgNO$_3$ solution, dissolve 0.85 g of AgNO$_3$ in enough distilled water to make 50 mL of solution.

2. To prepare 50 mL of 0.5 M Ca(NO$_3$)$_2$ solution, dissolve 4.1 g of Ca(NO$_3$)$_2$ in enough distilled water to make 50 mL of solution.

3. To prepare 50 mL of 0.2 M KBr solution, dissolve 1.2 g of KBr in enough distilled water to make 50 mL of solution.

For this experiment, wear safety goggles, gloves, and a lab apron to protect your eyes, hands, and clothing. If you get a chemical in your eyes, immediately flush the chemical out at the eyewash station while calling to your teacher. Know the location of the emergency lab shower and eyewash station and the procedure for using them.

Do not touch any chemicals. If you get a chemical on your skin or clothing, wash the chemical off at the sink while calling to your teacher. Make sure you carefully read the labels and follow the precautions on all containers of chemicals that you use. If there are no precautions stated on the label, ask your teacher what precautions you should follow. Do not taste any chemicals or items used in the laboratory. Never return leftovers to their original containers; take only small amounts to avoid wasting supplies.

Call your teacher in the event of a spill. Spills should be cleaned up promptly, according to your teacher's directions.

ChemFile

EXPERIMENT B5 **39**

EXPERIMENT B5 continued

4. To prepare 125 mL of 0.2 M KI, dissolve 4.1 g of KI in enough distilled water to make 125 mL of solution.

MATERIALS

5. To prepare 100 mL of 0.1 M NaCl solution, dissolve 0.6 g of NaCl in enough distilled water to make 100 mL of solution.

6. To prepare 125 mL of 0.1 M NaF solution, dissolve 0.5 g of NaF in enough distilled water to make 125 mL of solution.

PROCEDURE

7. To prepare 100 mL of 0.2 M $Na_2S_2O_3$ solution, dissolve 5.0 g of $Na_2S_2O_3 \cdot 5H_2O$ in enough distilled water to make 100 mL of solution.

8. To prepare 500 mL of 4 M aqueous NH_3, dilute 134 mL of concentrated ammonia water to a volume of 500 mL.

9. To prepare a 3% starch solution, dissolve 3.0 g of soluble starch in approximately 80 mL of water. Boil, and then dilute to 100 mL of solution.

10. If a mixture of two ions is used for an unknown, double the concentration of each original solution.

Required Precautions

- Safety goggles and a lab apron must be worn at all times.
- Read all safety precautions, and discuss them with your students.

Never put broken glass into a regular waste container. Broken glass should be disposed of properly in the broken-glass waste container.

- 0.1 M $AgNO_3$
- 0.1 M NaCl
- 0.1 M NaF
- 0.2 M KBr
- 0.2 M KI
- 0.2 M $Na_2S_2O_3$
- 0.5 M $Ca(NO_3)_2$
- 3% starch solution
- 4 M $NH_3(aq)$
- 5% NaOCl (commercial bleach)
- 24-well microplate
- thin-stemmed pipets, 12

1. Put 5 drops of 0.1 M NaF into each of four wells in row A, as shown in Figure A. Put 5 drops of 0.1 M NaCl into each of the wells in row B. Put 5 drops of 0.2 M KBr into each of the wells in row C and 5 drops of 0.2 M KI into each of the wells in Row D. Reserve rows E and F for unknown solutions.

2. Add 5 drops of 0.5 M $Ca(NO_3)_2$ solution to each of the four halide solutions in column 1. Record your observations in the data table.

3. Add 2 drops of 0.1 M $AgNO_3$ solution to each of the halides in columns 2 and 3. Record in your data table the colors of the precipitates formed.

4. Add 5 drops of 4 M $NH_3(aq)$ to the precipitates in column 2. Record your observations in the data table.

EXPERIMENT B5 continued

- In case of a spill, use a dampened cloth or paper towel (or more than one towel if necessary) to mop up the spill. Then rinse the towel in running water at the sink, wring it out until it is only damp, and put it in the trash. In the event of an ammonia spill, dilute the spill with water and then proceed as described.

Disposal

Provide a plastic dishpan to collect the rinsings for step 9. Combine all solutions. Add enough 0.2 M KI to precipitate all of the Ag^+ as AgI. Then dilute 100-fold and pour the mixture down the drain.

5. Add 5 drops of 0.2 M $Na_2S_2O_3$ solution to the precipitates in column 3. Record your observations in the data table.

6. To the halides in column 4, add 5 drops of starch solution and 1 drop of 5% bleach solution. Record your observations. Save the results of testing the four known halide solutions for comparison with the tests of the unknown solutions.

7. Obtain an unknown solution. Put 5 drops of the unknown in each of the four wells in row E. Add the reagents to each well as you did in steps **2–5**. Compare the results with those of the known halides in rows A–D. Record your findings in the data table, and identify the unknown.

8. Obtain an unknown solution containing a mixture of two halide ions. Place 5 drops of the unknown mixture in each of the four wells in row F. Add the reagents to each well as you did in steps **2–5**. Record your results. Compare the results with those of the known halides in rows A–D. Identify the halides in the double unknown solution.

Cleanup and Disposal

9. Rinse the microplate into a trough or dishpan provided by your teacher. Clean all apparatus and your lab station. Return equipment to its proper place. Dispose of chemicals and solutions in the containers designated by your teacher. Do not pour any chemicals down the drain or in the trash unless your teacher directs you to do so. Wash your hands thoroughly before you leave the lab and after all work is finished.

Data Table

Halide salts	Ca(NO₃)₂	AgNO₃	AgNO₃ + Na₂S₂O₃	NaOCl + Starch
NaF	white ppt	clear		no change
NaCl	clear	white ppt	dissolves completely	no change
KBr	clear	pale-yellow ppt	dissolves completely	no change
KI	clear	yellow-tan ppt	does not dissolve	turns blue
Single unknown	student observations	student observations	student observations	student observations
Double unknown	student observations	student observations	student observations	student observations

ChemFile

EXPERIMENT B5 continued

QUESTIONS

1. **Analyzing Data** Which procedure(s) confirm(s) the presence of (a) F⁻ ions, (b) Cl⁻ ions, (c) Br⁻ ions, (d) I⁻ ions?

   a. Procedure 2

   b. Procedures 3 and 4

   c. Procedures 3 and 5

   d. Procedure 6

2. **Inferring Conclusions** What generalizations can be made about silver halides?

   They are all insoluble except for NaF.

GENERAL CONCLUSIONS

1. **Relating Ideas** In nuclear explosions or accidents, iodine 131, a radioactive fission product, can become dispersed in the atmosphere. Eventually, the iodine isotope will fall onto the ground and be absorbed by plants. Explain how radiation from iodine-131 could become concentrated in the human body and cause a growth disorder.

   Cattle eat the plants and concentrate the radioactive iodine in their milk. If humans drink the milk, the radioactive iodine will concentrate in their thyroid gland, causing a possible growth disorder.

2. **Inferring Conclusions** Identify your unknown(s) and use your experimental evidence to support your identifications.

   Students' answers will vary.

Name _____

Date _____ Class _____

EXPERIMENT B6

Chemical Bonds

OBJECTIVES

Recommended time:
50 min

- **Compare** the melting points of eight solids.
- **Determine** the solubilities of the solids in water and in ethanol.
- **Determine** the conductivity of water solutions of the soluble solids.
- **Classify** the compounds into groups of ionic and covalent compounds.
- **Summarize** the properties of each group.

INTRODUCTION

Solution/Material Preparation

1. Use small test tubes for the solubility tests if microtitration plates are not available.

2. Limit the total amount of ethanol in the room to 250 mL, in three or more bottles each containing no more than 100 mL.

Chemical compounds are combinations of atoms held together by chemical bonds. These chemical bonds are of two basic types—ionic and covalent. Ionic bonds result when one or more electron from one atom or group of atoms is transferred to another atom. Positive and negative ions are created through the transfer. In covalent compounds no electrons are transferred; instead electrons are shared by the bonded atoms.

The physical properties of a substance, such as melting point, solubility, and conductivity, can be used to predict the type of bond that binds the atoms of the compound. In this experiment, you will test eight compounds to determine these properties. Your compiled data will enable you to classify the substances as either ionic or covalent compounds.

SAFETY

3. If you plan to use any conductivity tester other than the microtester, use only apparatus powered by a 9 volt or smaller battery.

4. Provide a dishpan into which students can rinse their microtitration plates.

Required Precautions

- Safety goggles and a lab apron must be worn at all times.
- Read all safety precautions, and discuss them with your students.

Always wear safety goggles and a lab apron to protect your eyes and clothing. If you get a chemical in your eyes, immediately flush the chemical out at the eyewash station while calling to your teacher. Know the location of the emergency lab shower and eyewash station and the procedure for using them.

Do not touch any chemicals. If you get a chemical on your skin or clothing, wash the chemical off at the sink while calling to your teacher. Make sure you carefully read the labels and follow the precautions on all containers of chemicals that you use. If there are no precautions stated on the label, ask your teacher what precautions you should follow. Do not taste any chemicals or items used in the laboratory. Never return leftovers to their original containers; take only small amounts to avoid wasting supplies.

Call your teacher in the event of a spill. Spills should be cleaned up promptly, according to your teacher's directions.

ChemFile   EXPERIMENT B6 **43**

EXPERIMENT B6 continued

- In case of a spill, use a dampened cloth or paper towel (or more than one towel if necessary) to mop up the spill. Then rinse the towel in running water at the sink, wring it out until it is only damp and put it in the trash.

When you use a candle, confine long hair and loose clothing. If your clothing catches on fire, WALK to the emergency lab shower, and use it to put out the fire. Do not heat glassware that is broken, chipped, or cracked. Use tongs or a hot mitt to handle heated glassware and other equipment because hot glassware and other equipment does not look hot.

Ethanol is flammable. Make sure there are no flames in the room before you use this chemical.

MATERIALS

- Ethanol is flammable. Insure that there are no flames or other sources of ignition present when ethanol is used.

- 24-well microplate
- calcium chloride
- candle
- citric acid
- conductivity tester
- ethanol
- iron ring
- phenyl salicylate
- potassium iodide
- ring stand
- sodium chloride
- sucrose
- tin can lid
- thin-stemmed pipets, 2

PROCEDURE

Techniques to Demonstrate

Set up the tin can lid and candle on the ring stand and show students the amount of solid to be used. If students have not used a conductivity tester before, show them how to test it using a metal fitting such as a faucet. Demonstrate how to measure the conductivity of a solution, and caution students that the leads must be separated when they are placed in the solution.

Disposal

Collect all solutions and pour down the drain. Collect all solids and put them into the trash. Collect all can lids and put them into the trash. They are too hard to clean off thoroughly and have sharp edges.

1. Before you begin, write a brief description of each of the six substances in your data table.

2. Place a can lid on an iron ring attached to a ring stand. Position the ring so that it is just above the tip of a candle flame, as shown in Figure A. Light the candle for a moment to check that you have the correct height.

3. Place a *few* crystals of sucrose, sodium chloride, phenyl salicylate, calcium chloride, citric acid, and potassium iodide in separate locations on the lid, as shown in Figure B. Do not allow the samples of crystals to touch. Draw a diagram that shows the position of each compound.

| FIGURE A | FIGURE B |

EXPERIMENT B6 continued

4. For this experiment, it is not necessary to have exact values for the melting point. The lid will continue to get hotter as it is heated, so the order of melting will give relative melting points. Light the candle and observe. Note the substance that melts first by writing a *1* in the Data Table. Record the order of melting for the other substances.

5. After 2 min, record an *n* in your data table for each substance that did not melt. Extinguish the candle flame. Allow the tin can lid to cool while you complete the remainder of the experiment.

6. Put a *few* crystals of each of the white solids in the top row of your microtitration plate. Repeat with the second row. Add 10 drops of water to each well in the top row. Record the solubility of each substance in your data table.

7. Add 10 drops of ethanol to each well in the second row of the microtitration plate. Record the solubility of each substance in your data table.

8. Test the conductivity of each water solution in the top row by dipping both electrodes into each well of the microtitration plate. Be sure to rinse the electrodes and dry them with a paper towel after each test. If the bulb of the conductivity apparatus lights up, the solution conducts electricity. Record your results in your data table.

Cleanup and Disposal

9. Clean the microplate by rinsing it with water into a pan provided by your teacher. If any wells are difficult to clean, use a cotton swab. Wash your hands thoroughly before you leave the lab and after all work is finished.

Data Table

Compound	Description	Melting point	Solubility in water	Solubility in ethanol	Solution conductivity
Calcium chloride	student answer	n	sol	insol	yes
Citric acid	student answer	2	insol	sol	no
Phenyl salicylate	student answer	1	insol	sol	no
Potassium iodide	student answer	n	sol	insol	yes
Sodium chloride	student answer	n	sol	insol	yes
Sucrose	student answer	decomposes	sol	sol	no

EXPERIMENT B6 continued

QUESTIONS

1. **Organizing Results** Group the white substances according to their properties.

   sodium chloride   sucrose

   potassium chloride   phenyl salicylate

   calcium chloride   citric acid

2. **Organizing Results** List the properties of each group.

   high melting point   low melting point

   conducting solutions   nonconducting solutions

   water soluble   ethanol soluble

   not soluble in ethanol   not soluble in water

GENERAL CONCLUSIONS

1. **Inferring Conclusions** Use your textbook and your experimental data to determine which of the groups consists of ionic compounds and which consists of covalent compounds.

   Sodium chloride, calcium chloride, and potassium iodide are ionic compounds; sucrose, phenyl salicylate, and citric acid are covalent compounds.

2. **Relating Ideas** Write a statement to summarize the properties of ionic compounds and another statement to summarize the properties of covalent compounds.

   Ionic compounds are generally soluble in water, they have high melting points, and their water solutions conduct electricity.

   Covalent compounds are generally soluble in ethanol, they have low melting points, and their water solutions do not conduct electricity.

EXPERIMENT B7

Conductivity as an Indicator of Bond Type

OBJECTIVES

Recommended time:
50 min

- **Observe** and **compare** conductivities of various substances.
- **Relate** conductivity to the type of bonds in a substance.
- **Infer** the identities of unknown substances.

INTRODUCTION

Materials (not in the student's list)

- LED conductivity tester
  1-$k\Omega$ resistor (1/4 W)
  LED (light-emitting diode)
  9-V battery with battery clips
  film canister with lid
  10-cm and 20-cm lengths of hook-up wire (22 GA, solid)
  3.5-cm length of straw or tubing to fit over LED's nail or ice-pick
  electrician's tape
  glue gun (optional)
- ten substances, 3–4 g of the solids or 3–4 mL of the liquids yellow chalk dust, (grind up sticks of chalk)

For a substance to conduct an electrical current, the substance must possess free-moving charged particles. These charged particles may be delocalized electrons, such as those found in substances that form metallic bonds. The particles may also be mobile ions, such as those found in dissolved (or molten) salts, or they may be ions formed by certain molecular substances having polar-covalent bonds that dissociate when dissolved in water.

Solid ionic substances that do *not* dissolve in water are usually considered to be *insulators*, substances that do not conduct electricity. Substances with nonpolar-covalent bonds are also insulators, as are substances with polar-covalent bonds that are not easily broken to form ions when they are dissolved in water.

The fact that a substance is a conductor in its pure form, or conducts only in solution, or is an insulator defines a physical property of the substance and provides clues about the inner structure and type of bonding found in the substance. In this experiment, you will use qualitative measures of conductivity to determine the identities of 10 unknown substances, identified as A, B, C, D, E, F, G, H, I, and J. Each letter corresponds to one of these substances: sodium sulfate, sugar, sulfur, tap water, deionized (or distilled) water, tin, tin(IV) oxide, graphite, yellow chalk dust, and charcoal. These substances are not necessarily listed in the correct order. You will use your conductivity-test results and other information to identify each unknown as one of the substances listed.

SAFETY

- charcoal (grind up chunks or use charcoal dust)
- graphite
- sugar
- sulfur
- Na$_2$SO$_4$
- tap water
- tin
- SnO$_2$
- water, distilled

Always wear safety goggles and a lab apron to protect eyes and clothing. If you get a chemical in your eyes, immediately flush the chemical out at the eyewash station while calling to your teacher. Know the location of the emergency lab shower and eyewash station and the procedure for using them.

Do not touch any chemicals. If you get a chemical on your skin or clothing, wash the chemical off at the sink while calling to your teacher. Make sure you carefully read the labels and follow the precautions on all containers of chemicals that you use. If there are no precautions stated on the label, ask your teacher what precautions you should follow. Do not taste any chemicals or items used in the laboratory. Never return leftovers to their original containers; take only small amounts to avoid wasting supplies.

EXPERIMENT B7 continued

Call your teacher in the event of a spill. Spills should be cleaned up promptly, according to your teacher's directions.

MATERIALS

Optional for General Conclusions 2

- 1.0 M HCl solution, 3–4 mL
- iron nail
- NaCl, 3–4 g
- wax, 5 g

- dropper bottle of distilled water
- glass stirring rod, small, or toothpicks
- hand-held conductivity tester
- labeled containers, 10, each with a different unknown substance
- paper towels
- spatula
- spot plate
- thin-stemmed pipet

PROCEDURE

Solution/Material Preparation

1. Assign a letter from A through J to each of the 10 materials, and place them in containers that can be reused from year to year. Keep track of which letter corresponds to which material so that you can check students' work.

2. LED conductivity testers may be purchased from Lab Aids or made according to the directions on the next page. A conductivity tester that emits a tone rather than a light is also available from Lab Aids. Some students find it easier to distinguish small differences in conductivity by observing differences in tone rather than differences in intensity of light.

3. To prepare 50 mL of optional 1.0 M HCl solution for General Conclusions **2**, observe the required safety precautions. Add 4.3 mL of concentrated HCl, with stirring, to enough water to make 50 mL of solution.

1. Examine each of the 10 unknowns and record in your lab notebook your observations about their state (solid or liquid), color, texture, and other qualities.

2. Test your conductivity apparatus to make sure it lights up when the two wires or leads touch different parts of a conducting substance. Your teacher will tell you what conducting substance to use.

3. Using a clean spatula for each solid and a clean pipet for each liquid, deliver small samples of each unknown into separate wells of a spot plate. **Keep track of which unknown is in which well. The spatula should be wiped with a clean paper towel between each use to prevent contamination. Similarly, the pipet should be wiped with a clean paper towel and rinsed with distilled water between each use.** Put any water used for rinsing into the sink.

4. Place the wire leads in good contact with the first unknown sample, as shown in Figure A. Note whether or not it conducts, as indicated by the light on the conductivity tester. (It may help to move the tester around a bit while watching the light.) If the two leads on the tester touch each other, the light will automatically go on. Do not mistake this for a positive test. Some results may be faint and require careful observations. Most pure substances are nonconductors, so don't expect a majority of your substances to conduct. Record all observations in your lab notebook.

EXPERIMENT B7 continued

Conductivity Tester Assembly

See page 52 for a diagram of this apparatus.

1. Strip 1.5 cm of insulation from the ends of the hook-up wires and from the battery clip leads. Cut the resistor leads down to half of their length.

2. Poke two holes near the edge of the canister cap about as far apart as the two LED leads are. Insert the leads through the holes, and spread them apart slightly.

5. Repeat step 4 for the other unknown samples. *Be sure to thoroughly wipe the leads of the conductivity probe with a clean paper towel between each test.* For those substances that do conduct, compare the conductivities and rate them on a scale from 0 to 4, with 4 as the best conductor. Record your observations and ratings in your lab notebook.

6. Using the pipet, add 10 drops of distilled water to each of the 10 unknowns. *Hold the tip of the pipet 1 to 2 cm above the unknowns.* Use a small glass stirring rod or toothpicks to stir each mixture. *Be sure to wipe the stirring rod or toothpick thoroughly with a clean paper towel after each use.*

7. Observe each of the wells, and record in your lab notebook any observations about the dissolving or mixing of the unknowns.

8. Test the conductivity of each mixture, again wiping the leads thoroughly between each test. Record your results.

Cleanup and Disposal

9. Clean all apparatus and your lab station. Return equipment to its proper place. Dispose of chemicals and solutions in the containers designated by your teacher. Do not pour any chemicals down the drain or in the trash unless your teacher directs you to do so. Wash your hands thoroughly before you leave the lab and after all work is finished.

QUESTIONS

3. Connect the circuit as shown in the illustration. **Be sure the *red* battery clip lead is connected to the *longer* LED lead.** The black battery clip lead should connect to the resistor, and the other end of the resistor should connect to the 10 cm length of wire. The shorter LED lead should be connected to the 20 cm wire. Bend the leads up and tape them against the inside of the lid so that they are not touching each other and there are no exposed wires.

4. Wrap tape around the resistor and its exposed wires.

5. Poke two holes in the side of a film canister about 1 cm apart, near the bottom of the film canister. Insert the hook-up wires through these holes.

1. **Organizing Ideas** Which types of bonding are likely to be involved in the unknowns that were conductors in their pure form? Explain what free-moving charged particles are available in these types of bonds.

   Metallic bonding and other bonds involving free electrons, such as the bonds between graphite sheets, are the types of bonding that should conduct in pure form. The free-moving electrons are the charged particles that carry the current.

2. **Organizing Ideas** Which types of bonding are likely be involved in the unknowns that were not conductors in their pure form but conducted electricity when they were mixed with water? Explain what free-moving charged particles are available in these types of bonds.

   Ionic substances that dissolve in water and polar-covalent substances that form ions when dissolved in water are likely to conduct electricity when mixed with water. The ions themselves serve as free-moving charged particles to carry the current.

EXPERIMENT B7 continued

6. As shown in the illustration, connect the battery and resistor, and place them in the canister, snapping the lid shut. Adjust the protruding hook-up wires so they extend out about 5 cm. (Optional: place drops of hot-melt glue around the LED and the hook-up wires so that the canister is waterproof.)

7. Wrap electrician's tape around the straw, and place it over the LED to provide a "dark tunnel" through which the LED may be viewed for testing weak electrolytes.

3. Organizing ideas Which types of bonding are likely to be involved in the unknowns that were not conductors in their pure form or when they were mixed with water? Explain why there were no free-moving charged particles available in these types of bonds.

The nonpolar-covalent substances and those ionic substances that do not dissolve in water are likely to be insulators in their pure form. The covalent bonds in nonpolar molecules hold the electrons too firmly for them to carry the current. In ionic substances that do not dissolve in water, the ions are firmly bound in a lattice, and are not available to move freely and carry the current.

4. Organizing Data and Relating Ideas Fill in the table below by using the *Useful Information* that follows to determine which type of bonding is likely to be present in each of the substances listed and whether the substances are likely to conduct electricity. Predict which will be the best conductor in pure form and dissolved in water, which will conduct only marginally, and which will not conduct at all.

Substance	Type of bonding	Conductivity	Conductivity in H_2O	Matches unknown
Chalk	ionic	no	no	
Charcoal	nonpolar-covalent	no	not soluble	
Graphite	nonpolar-covalent with some free electrons	slight	not soluble	
Sugar	nonpolar-covalent	no	no, because it does not form ions	
Sulfur	nonpolar-covalent	no	no	
Sodium sulfate	ionic	no	yes	
Tin	metallic	yes	not soluble	
Tin oxide	ionic	no	not soluble	
Pure H_2O	polar-covalent	no		
Tap water	polar-covalent	slight		

EXPERIMENT B7 continued

USEFUL INFORMATION

Required Precautions

- Safety goggles and a lab apron must be worn at all times.
- Read all safety precautions, and discuss them with your students.

GENERAL CONCLUSIONS

- In case of a spill, use a dampened cloth or paper towel (or more than one towel if necessary) to mop up the spill. Then rinse the towel in running water at the sink, wring it out until it is only damp and put it in the trash.

Techniques to Demonstrate

Show students how to use the conductivity tester, and suggest materials in the lab that they can use to test the apparatus before beginning the experiment, for example, the laboratory faucets.

You may wish to emphasize the role of hypotheses in scientific work by having students try to narrow the unknowns before they begin the experiment. Have them use the information provided in *Useful Information* in Calculations item **4** to make predictions about conductivities before they perform the conductivity tests. This approach, however, will take more time.

- Chalk is mostly $CaCO_3$ and dissolves only slightly in water.
- Sulfur is a yellow molecular substance, arranged in S_8 ring molecules.
- Sodium sulfate is white and dissolves in water.
- Tin is a metal.
- Tin oxide is white and does not dissolve in water.
- Charcoal is a form of carbon with the atoms bonded to each other in a randomly organized way.
- Graphite is a form of carbon. The atoms are bonded to each other in systematic sheets with some free electrons holding the sheets together.
- Sugar is a molecular substance that dissolves in water.
- Distilled or deionized water can be considered pure H_2O.
- Tap water contains many dissolved impurities that are molecular and ionic.

1. In the fifth column of your table, write the letter of the unknown that corresponds to the substance in each row. Explain your reasons for making each identification.

   Students' answers will vary, depending on the assignment of letter designations to the unknowns. The following are some possible rationales for their identifications:

   Chalk is an ionic compound that does not dissolve in water, so it will not conduct. Charcoal is a molecule with nonpolar-covalent bonds that will not conduct. Graphite consists mostly of nonpolar-covalent bonds, but it also has some free electrons between bonding layers, so it will conduct slightly. Sugar is a molecule with nonpolar-covalent bonds that will not conduct electricity even when dissolved in water. Sulfur is a molecule with nonpolar-covalent bonds that will not conduct. Sodium sulfate is an ionic compound that will conduct after it is dissolved in water. Tin has metallic bonds and will conduct electricity. Tin oxide is an ionic compound that does not conduct in solid form and does not dissolve in water. Distilled or deionized water has polar-covalent bonds that do not dissociate appreciably, so it will not conduct.

2. **Designing Experiments and Predicting Outcomes** Predict the conductivities of each of the materials listed below, and give reasons for your predictions. How would the conductivity of each material be affected if water

EXPERIMENT B7 continued

Disposal

All solutions can be washed down the sink with plenty of water, provided your school drains are connected to a sanitary sewer system with a treatment plant. The solid precipitates should be dried and buried in a landfill that is approved for chemical disposal. You may want to consider drying and reusing the materials that are not soluble in water. In addition, the tin has value as scrap metal, so you may want to recycle it instead of discarding it. The microplates or spot plates can be wiped with a clean paper towel, rinsed with tap water, rinsed with distilled water, and allowed to dry.

were added to it? If your teacher approves, test your predictions for all the materials except liquid bromine, which is too dangerous to work with.

a. an iron nail
b. liquid bromine, Br_2
c. wax (a random arrangement of molecules with long, nonpolar chains of carbon and hydrogen atoms)
d. a dilute hydrochloric acid solution, $HCl(aq)$
e. table salt, NaCl

An iron nail should conduct, because it contains metallic bonds with free electrons. Liquid bromine is a nonpolar-covalently bonded molecule, so it should not conduct. Wax is a nonpolar-covalently bonded molecule that should not conduct. Hydrochloric acid is a polar-covalent molecule that dissociates in water, so it will conduct. If water is added, the conductivity of HCl will decrease as the concentration decreases. Table salt will not conduct in the solid form, but because it is an ionic compound, it will conduct when dissolved in water.

Name _____

Date _____ Class _____

EXPERIMENT B8

Tests for Iron(II) and Iron(III)

OBJECTIVES

Recommended time: 50 min

- **Observe** tests of known solutions containing iron(II) or iron(III) ions.
- **Compare** results for the two ions and infer conclusions.
- **Design** a procedure for identifying the two ions in one solution.

INTRODUCTION

Solution/Material Preparation

1. The five solutions should be placed in the thin-stemmed pipets.
2. In preparing the solutions, care should be taken that the solid potassium ferrocyanide and potassium ferricyanide are not heated or come into contact with concentrated acids.

In this experiment, the complex hexacyanoferrate(II) ion (*ferro*cyanide), $Fe(CN)_6^{4-}$, and the hexacyanoferrate(III) ion (*ferri*cyanide), $Fe(CN)_6^{3-}$, will be used in identification tests for Fe^{2+} and Fe^{3+} ions. The charges on the two complex ions clearly indicate the difference in the oxidation state of the iron present in each. The (CN) group in each complex ion has a charge of -1. Thus, iron(II) is present in the *ferro*cyanide ion, $[Fe^{2+}(CN^-)_6]^{4-}$. Iron(III) is present in the *ferri*cyanide ion group, $[Fe^{3+}(CN^-)_6]^{3-}$. A deep blue precipitate results when either complex ion combines with iron in a different oxidation state from that present in the complex. The deep blue color of the precipitate is caused by the presence of iron in both oxidation states. The color provides a means of identifying either iron ion. If the deep blue precipitate is formed on addition of the $[Fe^{2+}(CN^-)_6]^{4-}$ complex, the iron ion responsible must be the iron(III) ion. Similarly, a deep blue precipitate formed with the $[Fe^{3+}(CN^-)_6]^{3-}$ complex indicates the presence of the iron(II) ion.

Both of the deep blue precipitates are known to have the same composition. The potassium salt of the complex ion has the formula $KFeFe(CN)_6 \cdot H_2O$.

The thiocyanate ion, SCN^-, provides a test for confirming the presence of Fe^{3+} ion. The soluble $FeSCN^{2+}$ complex imparts a blood red color to the solution.

SAFETY

3. The iron(II) ammonium sulfate solution must be freshly prepared because it will oxidize. To prepare 50 mL of 0.2 M iron(II) ammonium sulfate, dissolve 3.9 g of $Fe(NH_4)_2(SO_4)_2 \cdot 6H_2O$ in enough water to make 50 mL of solution.

Always wear safety goggles and a lab apron to protect your eyes and clothing. If you get a chemical in your eyes, immediately flush the chemical out at the eyewash station while calling to your teacher. Know the location of the emergency lab shower and eyewash station and the procedure for using them.

Do not touch any chemicals. If you get a chemical on your skin or clothing, wash the chemical off at the sink while calling to your teacher. Make sure you carefully read the labels and follow the precautions on all containers of chemicals that you use. If there are no precautions stated on the label, ask your teacher what precautions you should follow. Do not taste any chemicals or items used in the laboratory. Never return leftovers to their original containers; take only small amounts to avoid wasting supplies.

EXPERIMENT B8 continued

Call your teacher in the event of a spill. Spills should be cleaned up promptly, according to your teacher's directions.

MATERIALS

- 0.1 M FeCl$_3$
- 0.1 M K$_3$Fe(CN)$_6$
- 0.1 M K$_4$Fe(CN)$_6$
- 0.2 M Fe(NH$_4$)$_2$(SO$_4$)$_2$
- 0.2 M KSCN
- plastic wrap, 8 cm × 30 cm
- thin-stemmed pipets, 5

PROCEDURE

4. To prepare 50 L of 0.1 M iron(III) chloride, observe the required safety precautions. Add 1 mL of concentrated HCl, with stirring, to approximately 40 mL of water. Dissolve 1.35 g of FeCl$_3$ • 6H$_2$O in the acidified water and then dilute to 50 mL.

5. To prepare 100 mL of 0.1 M potassium hexacyanoferrate(III), dissolve 3.4 g of K$_3$Fe(CN)$_6$ in enough water to make 100 mL of solution.

6. To prepare 100 mL of 0.1 M potassium hexacyanoferrate(II), dissolve 4.2 g of K$_4$Fe(CN)$_6$ • 3H$_2$O in enough water to make 100 mL of solution.

7. To prepare 100 mL of 0.1 M potassium thiocyanate, dissolve 1.9 g KSCN in enough water to make 100 mL of solution.

Required Precautions

- Safety goggles and a lab apron must be worn at all times.
- Read all safety precautions, and discuss them with your students.

1. Place the plastic-wrap rectangle on a white sheet of paper.

2. Along the top of the plastic wrap, place 5 drops of a freshly prepared iron(II) ammonium sulfate solution in each of the three locations shown in Figure A.

3. Along the bottom of the plastic wrap, place 5 drops of a freshly prepared iron(III) chloride solution, as shown in Figure A.

FIGURE A

4. Add 1 drop of 0.1 M K$_4$Fe(CN)$_6$ solution to the first sample of iron(II) ions at the top of the plastic wrap and 1 drop to the first sample of iron(III) ions at the bottom. Record your observations in the Data Table.

5. Add 1 drop of 0.1 M KSCN solution to the second sample of iron(II) ions and 1 drop to the second sample of iron(III) ions. Record your observations in the Data Table.

6. Add 1 drop of 0.1 M K$_3$Fe(CN)$_6$ solution to the third sample of iron(II) ions and 1 drop to the third sample of iron(III) ions. Record your observations.

EXPERIMENT B8 continued

- In case of a spill, use a dampened cloth or paper towel (or more than one towel if necessary) to mop up the spill. Then rinse the towel in running water at the sink, wring it out until it is only damp and put it in the trash. In the event of a spill of acidic FeCl₃, dilute the spill first with water and then proceed as described.
- Wear goggles, face shield, impermeable gloves, and a lab apron when you prepare the iron(III) chloride solution containing HCl. Work in a hood known to be in operating condition, with another person present nearby to call for help in case of an emergency. Be sure you are a 30 s walk from a safety shower and eyewash station known to be in working condition.

QUESTIONS

Disposal

Combine all solutions and precipitates. Filter. To the filtrate, add 0.1 M potassium hexacyanoferrate(III) solution until no more precipitate forms. Filter. To the filtrate, add 0.1 M potassium hexacyanoferrate(II) solution until no more precipitate forms. Filter. Adjust the pH of the filtrate to be within 5 to 9 by adding 1.0 M NaOH solution if necessary. Then pour down the drain. Combine all precipitates and put in the trash after drying.

Cleanup and Disposal

7. Clean all apparatus and your lab station. Return equipment to its proper place. Dispose of chemicals and solutions in the containers designated by your teacher. Do not pour any chemicals down the drain or in the trash unless your teacher directs you to do so. Wash your hands thoroughly before you leave the lab and after all work is finished.

Data Table

Iron ion	Hexacyanoferrate (II) ion $[Fe^{2+}(CN^-)_6]^{4-}$	Hexacyanoferrate (III) ion $[Fe^{3+}(CN^-)_6]^{3-}$	Thiocyanate ion SCN^-
Fe^{2+}	white ppt	dark blue ppt	no visible result
Fe^{3+}	dark blue ppt	brown solution	blood red solution
Observations in step 4	With iron(II), no precipitate formed, but with iron(III), a deep blue precipitate formed.		
Observations in step 5	With iron(III), a deep red color appeared; there was no reaction with iron(II) ions.		
Observations in step 6	A blue precipitate formed with the iron(II) ion; there was no precipitate with iron(III).		

1. Organizing Ideas Explain specifically how you would make a conclusive test for an iron(III) salt.

Add $K_4Fe(CN)_6$ to the solution. The formation of a deep blue precipitate identifies the iron(III) ion. Confirm by adding KSCN to a second test solution. The formation of a blood red solution confirms the presence of an iron(III) salt.

2. Organizing Ideas Which test for iron(II) ions is conclusive?

The addition of $K_3Fe(CN)_6$ to a solution causes the formation of a dark blue precipitate which indicates that iron(II) ions are present.

ChemFile

EXPERIMENT B8 continued

3. **Relating Ideas** When iron(II) ammonium sulfate was mixed with the $[Fe^{2+}(CN^-)_6]^{4-}$ ion, the precipitate was initially white but turned blue upon exposure to air. What happened to the iron(II) ion when the precipitate turned blue?

   Iron(II) was oxidized to iron(III) in the presence of air.

GENERAL CONCLUSIONS

1. **Designing Experiments** Suppose you have a solution containing both an iron(II) salt and an iron(III) salt. How would you proceed to identify both Fe^{2+} and Fe^{3+} in this solution?

   $K_3Fe(CN)_6$ could be added to one sample of the solution to detect

   the iron(II) ion, and $K_4Fe(CN)_6$ could be added to another sample

   of the solution to identify the iron(III) ion.

2. **Relating Ideas** Blueprint paper can be made by soaking paper in a brown solution of $[Fe^{3+}(CN^-)_6]^{3-}$ and iron(III) ammonium citrate. Wherever the paper is exposed to bright light, the paper turns blue. Explain why this happens.

   The bright light changes the iron(III) ion to iron(II), which then

   precipitates with the ferricyanide as Prussian blue. This deep blue

   compound is insoluble in water.

Name _____

Date _____ Class _____

EXPERIMENT B9

Colored Precipitates

OBJECTIVES

Recommended time: 30 min plus more time for activities in General Conclusions

- **Observe** replacement reactions in which precipitates are formed.
- **Compare** chemical and physical properties of substances.
- **Relate** observations to the identification of unknown solutions.
- **Infer** a conclusion from experimental data.

INTRODUCTION

Optional Materials for General Conclusions items 3 and 4

- Filter paper, 7 pieces
- Funnel
- Balance
- pH paper sensitive in the range 12–14

Binary ionic compounds exhibit a variety of properties. Some dissolve easily in water, while others are insoluble. Some are brightly colored compounds; many others are white. You can use characteristic properties of known substances to determine the identity of an unknown by comparing the properties of a known substance with those of the unknown.

In this experiment, you will carry out some double-replacement reactions and observe the characteristic colors of precipitates. You will use your experimental data to identify unknown substances.

SAFETY

Solution/Material Preparation

1. To prepare 25 mL of 0.1 M cobalt(II) chloride solution, dissolve 0.60 g of $CoCl_2 \cdot 6H_2O$ in enough distilled water to make 25 mL of solution.

2. To prepare 25 mL of 0.1 M copper(II) chloride solution, dissolve 0.43 g of $CuCl_2 \cdot 2H_2O$ in enough distilled water to make 25 mL of solution.

Always wear safety goggles and a lab apron to protect your eyes and clothing. If you get a chemical in your eyes, immediately flush the chemical out at the eyewash station while calling to your teacher. Know the location of the emergency lab shower and eyewash station and the procedure for using them.

Do not touch any chemicals. If you get a chemical on your skin or clothing, wash the chemical off at the sink while calling to your teacher. Make sure you carefully read the labels and follow the precautions on all containers of chemicals that you use. If there are no precautions stated on the label, ask your teacher what precautions you should follow. Do not taste any chemicals or items used in the laboratory. Never return leftovers to their original containers; take only small amounts to avoid wasting supplies.

Call your teacher in the event of a spill. Spills should be cleaned up promptly, according to your teacher's directions.

MATERIALS

- 0.1 M $CoCl_2$
- 0.1 M $CuCl_2$
- 0.1 M $FeCl_3$
- 0.1 M NaOH
- 0.1 M $NiCl_2$
- 8-well flat-bottom strip or small test tubes
- mystery solutions 1, 2, and 3
- thin-stemmed pipets, 8

EXPERIMENT B9 continued

PROCEDURE

3. To prepare 25 mL of 0.1 M iron(III) chloride solution, dissolve 0.68 g of $FeCl_3 \cdot 6H_2O$ in enough distilled water to make 25 mL of solution.

4. To prepare 25 mL of 0.1 M nickel(II) chloride solution, dissolve 0.60 g of $NiCl_2 \cdot 6H_2O$ in enough distilled water to make 25 mL of solution.

5. To prepare 250 mL of 0.1 M sodium hydroxide solution, dissolve 1.0 g of NaOH in enough distilled water to make 250 mL of solution.

6. To prepare 25 mL of mystery solution *1* (0.1 M iron(III) sulfate solution), dissolve 1.40 g of $Fe_3(SO_4)_2 \cdot 9H_2O$ in enough distilled water to make 25 mL of solution.

7. To prepare 25 mL of mystery solution *2* (0.1 M nickel(II) sulfate solution), dissolve 1.32 g of $NiSO_4 \cdot 6H_2O$ in enough distilled water to make 25 mL of solution.

8. To prepare 25 mL of mystery solution *3* (0.2 M copper(II) sulfate solution), dissolve 1.25 g of $CuSO_4 \cdot 5H_2O$ in enough distilled water to make 25 mL of solution.

9. Thin-stemmed pipets containing the reagents may be shared by an entire class.

Required Precautions

- Safety goggles and a lab apron must be worn at all times.
- Read all safety precautions, and discuss them with your students.

1. Label a sheet of paper to identify the solutions, as shown in Figure A. Place it under the 8-well strip.

FIGURE A

(Figure shows a hand with a pipet adding Cobalt chloride solution to the first of 8 wells on a strip, with a sheet of paper labeled: Cobalt, Copper, Iron, Nickel, Mystery 1, Mystery 2, Mystery 3. 20 drops of NaOH solution labeled above the wells.)

2. With a thin-stemmed pipet, place 20 drops of sodium hydroxide in the first seven of the eight wells in the strip. (If you are using test tubes, use 40 drops.) *For best results, try to keep all the drops the same size.*

3. To the first well, add 5 drops of cobalt(II) chloride solution. (If using test tubes, add 10 drops of each metal chloride to each tube.) To the second, add 5 drops of copper(II) chloride solution. To the third, add 5 drops of iron(III) chloride. To the fourth, add 5 drops of nickel(II) chloride. Observe what happens. Record your observations and the physical properties of the substances formed in the Data Table.

4. To the fifth well, add 5 drops of mystery solution *1*. (If using test tubes, add 10 drops of each of the mystery solutions.) To the sixth well, add 5 drops of mystery solution *2*. To the seventh well, add 5 drops of mystery solution *3*. Observe what happens. Record your observations about the kinds of reactions you see and the physical properties of the substances formed.

Cleanup and Disposal

5. Clean all apparatus and your lab station. Return equipment to its proper place. Dispose of chemicals and solutions in the containers designated by your teacher. Do not pour any chemicals down the drain or in the trash unless your teacher directs you to do so. Wash your hands thoroughly before you leave the lab and after all work is finished.

EXPERIMENT B9 continued

Data Table

Well Number	1	2	3	4	5	6	7
Label	CoCl$_2$	CuCl$_2$	FeCl$_3$	NiCl$_2$	Mystery solution 1	Mystery solution 2	Mystery solution 3
Observations in step 3	Rose-red ppt	Blue ppt	Pale-green ppt	Green ppt	×	×	×
Observations in step 4	×	×	×	×			

QUESTIONS

- In case of a spill, use a dampened cloth or paper towel (or more than one towel if necessary) to mop up the spill. Then rinse the towel in running water at the sink, wring it out until it is only damp, and put it in the trash.

Pre-lab Discussion

Review double replacement reactions. If necessary, discuss the writing of net ionic equations presented in item **2**.

Disposal

Combine all solutions and precipitates. With stirring, add an excess of 0.1 M NaOH sufficient to raise the pH to approximately 10. Then, with stirring, slowly add 0.1 M HCl to reduce the pH to 8; do not allow the pH to become less than 8. Allow the mixture to dry; then wrap it in newspaper and put it in the trash.

1. **Organizing Ideas** The reactions in this lab are all double-replacement reactions. Write chemical equations for the reactions occurring in the first four mixtures.

    CoCl$_2$(aq) + 2NaOH(aq) → Co(OH)$_2$(s) + 2NaCl(aq)

    CuCl$_2$(aq) + 2NaOH(aq) → Cu(OH)$_2$(s) + 2NaCl(aq)

    FeCl$_3$(aq) + 3NaOH(aq) → Fe(OH)$_3$(s)* + 3NaCl(aq)

    NiCl$_2$(aq) + 2NaOH(aq) → Ni(OH)$_2$(s) + 2NaCl(aq)

    * To be accurate, the substance formed is FeOOH, or FeOOH·H$_2$O (FeO$_3$H$_3$).

2. **Organizing Ideas** A net ionic equation is one that includes only those dissolved ions that undergo a reaction. For example, in writing a net ionic equation for the reaction of HCl with NaOH, the following steps are taken.
    - Write the balanced equation showing the complete formulas for the reactants and products.

        HCl(aq) + NaOH(aq) → NaCl(aq) + H$_2$O(l)

    - Rewrite the equation with all ions separated or dissociated. Formulas for solid precipitates, gases, or liquids should be written as undissociated compounds.

        H$^+$(aq) + Cl$^-$(aq) + Na$^+$(aq) + OH$^-$(aq) → Na$^+$(aq) + Cl$^-$(aq) + H$_2$O(l)

    - Cross out any ions that appear in the same form on both sides of the equation (in this case, Na$^+$ and Cl$^-$).

        H$^+$(aq) + ~~Cl$^-$(aq)~~ + ~~Na$^+$(aq)~~ + OH$^-$(aq) → ~~Na$^+$(aq)~~ + ~~Cl$^-$(aq)~~ + H$_2$O(l)

    - Rewrite the equation omitting any ions that were crossed out.

        H$^+$(aq) + OH$^-$(aq) → H$_2$O(l)

ChemFile

EXPERIMENT B9 continued

This is the net ionic equation that includes only those ions undergoing a chemical change. Write a net ionic equation for the reactions that occurred in the first four mixtures.

$Co^{2+}(aq) + 2OH^-(aq) \rightarrow Co(OH)_2(s)$

$Cu^{2+}(aq) + 2OH^-(aq) \rightarrow Cu(OH)_2(s)$

$Fe^{3+}(aq) + 3OH^-(aq) \rightarrow Fe(OH)_3(s)$

$Ni^{2+}(aq) + 2OH^-(aq) \rightarrow Ni(OH)_2(s)$

3. **Analyzing Results** Would the reactions have been different if potassium hydroxide (KOH) had been used instead of sodium hydroxide? Explain your answer. Hint: Write the net ionic equation using KOH(*aq*) instead of NaOH(*aq*).

The net ionic equations are the same as those in item 2 for NaOH.

The Na$^+$ and K$^+$ ions play no role in the reactions.

4. **Inferring Conclusions** Which of the net ionic equations that you wrote in item 2 describe the reactions of the three mystery solutions? Write the equations, and explain the reasons you chose them.

Mystery solution 1: $Fe^{3+}(aq) + 3OH^-(aq) \rightarrow Fe(OH)_3(s)$

Mystery solution 2: $Ni^{2+}(aq) + 2OH^-(aq) \rightarrow Ni(OH)_2(s)$

Mystery solution 3: $Cu^{2+}(aq) + 2OH^-(aq) \rightarrow Cu(OH)_2(s)$

The identifications are based on the colors of the precipitates.

5. **Relating Ideas** All the reactants in these reactions are solutions of ionic compounds. Do you think the characteristic colors of the reactants and the products are caused by the cations or the anions in the compounds? Explain your reasoning.

Students' answers will vary but should note that the color of each

precipitate is caused by the positive ion (cobalt, copper, iron, or

nickel). The color can't be caused by the negative hydroxide ion

because all the compounds formed were hydroxides, yet each had

a different color.

EXPERIMENT B9 continued

6. **Relating Ideas** How many drops of NaOH are needed to react completely with 5 drops of each of the metal chloride solutions, if all the drops are the same size? (Hint: use the balanced chemical equations and recall that the concentrations of all the reactants are the same.)

If the drops are the same size and the concentrations are the same, inspection of the balanced equations reveals that the cobalt, copper, and nickel solutions require twice as much NaOH solution (10 drops), and the iron solution requires three times as much NaOH solution (15 drops).

7. **Predicting Outcomes** Use your answer to item 6 to decide which reactant, NaOH or metal chloride, would be left over in each well after the reaction? How much of the excess reactant would be present?

Because 20 drops of NaOH were used, there should be 5 drops of NaOH left unreacted after all the iron chloride has reacted and 10 drops of NaOH left unreacted after the other metal chlorides have reacted.

GENERAL CONCLUSIONS

1. **Analyzing Results** Can you tell if the mystery solutions had concentrations of positive ions that were greater than, less than, or the same as the concentrations of the known solutions you tested? Explain your reasoning.

Students' answers will vary, but if they were able to tell that the mystery solutions had a higher concentration of the metallic cations, the evidence cited should include the formation of a larger amount of precipitate when the mystery solutions reacted.

EXPERIMENT B9 continued

2. **Designing Experiments** Can you think of a way to use the products of these reactions to make an estimate of the concentration of the mystery solutions? Write a procedure for making this determination. If your teacher approves of your suggestions, test your procedure.

Students' answers will vary, but students may suggest filtering the precipitates, drying them, and measuring their mass. The mass of metal hydroxide formed by one of the known solutions will be proportional to its concentration, and the same proportion will hold for the mystery solution containing that metal chloride. Be sure that procedures students propose are carefully planned and safe.

3. **Designing Experiments** The 0.1 M NaOH solution used as a reactant in this reaction has a pH of 13.0. The mathematical relationship of pH to OH^- concentration in moles per liter is given by the following equation.

$$pH = 14.00 + \log [OH^-]$$

Describe how you could use a pH meter to estimate the concentration of the positive ions in the mystery solutions.

Students may suggest using a pH meter to accurately measure the pH of the solution remaining after the reaction. By comparing the pH measurements of the reacted mystery solutions with those of the known standards and calculating the concentration of the OH^- ions remaining, they could estimate the number of OH^- ions that reacted and the number of positive ions associated with those OH^- ions. Be sure that student procedures are carefully planned and are safe.

EXPERIMENT B10

Simple Qualitative Analysis

OBJECTIVES

Recommended time: 50 min

- **Observe** qualitative tests using known ionic compounds.
- **Decide** which qualitative test to use in identifying an unknown ionic compound.
- **Describe** the chemistry of common ionic compounds.

INTRODUCTION

Solution/Material Preparation

1. To prepare 200 mL of 1.0 M HCl, observe the required safety precautions. Slowly, and with stirring, add 18 mL of concentrated HCl to enough distilled water to make 200 mL of solution.

2. To prepare 250 mL of 1.0 M NaOH, observe the required safety precautions. Slowly, and with stirring, dissolve 10 g of NaOH in enough distilled water to make 250 mL of solution.

If an unknown sample is one of a limited number of possible compounds, a simple test often can determine its identity. For example, a flame test can distinguish between KCl and NaNO$_3$. A drop of potassium hexacyanoferrate(III) solution can tell you whether FeCl$_2$ or FeCl$_3$ is present. There are hundreds of simple qualitative tests such as these to distinguish among a few possibilities.

Qualitative tests are important to the forensic chemist, one who is interested in solving crimes, but they have been largely replaced by instrumental analysis, which is fast and requires a very small sample. For example, instrumental analysis can detect as little as 1×10^{-11} g of 9-tetrahydrocannabinol (one of the active ingredients in marijuana) in 1 mL of blood plasma. Unfortunately, this kind of instrumentation (a gas chromatograph connected to a mass spectrograph) is expensive, so simple chemical tests are still often useful. A drop of hydrochloric acid is enough to allow the police department's forensic chemist to distinguish between a bag of cocaine and a bag of baking soda.

In this experiment you will identify the contents of a number of vials. The substance in each vial is one of the two compounds listed on the label. To decide which compound is present, you will make a few simple tests.

SAFETY

3. To prepare 500 mL of 0.1 M potassium permanganate solution, dissolve 8 g of KMnO$_4$ in enough distilled water to make 500 mL of solution.

4. To prepare 500 mL of 0.1 M potassium hexacyanoferrate(III), dissolve 17 g of K$_3$Fe(CN)$_6$ in enough water to make 500 mL of solution.

For this experiment, wear safety goggles, gloves, and a lab apron to protect eyes, hands, and clothing. If you get a chemical in your eyes, immediately flush the chemical out at the eyewash station while calling to your teacher. Know the location of the emergency lab shower and eyewash station and the procedure for using them.

Do not touch any chemicals. If you get a chemical on your skin or clothing, wash the chemical off at the sink while calling to your teacher. Make sure you carefully read the labels and follow the precautions on all containers of chemicals that you use. If there are no precautions stated on the label, ask your teacher what precautions you should follow. Do not taste any chemicals or items used in the laboratory. Never return leftovers to their original containers; take only small amounts to avoid wasting supplies.

EXPERIMENT B10 continued

5. To prepare 500 mL of 0.1 M potassium hexacyanoferrate(II), dissolve 21 g of $K_4Fe(CN)_6 \cdot 3H_2O$ in enough water to make 500 mL of solution.

6. To prepare the unknowns, place 15 g of *one* compound from each of the following pairs into each vial and label the vial with the names of both compounds in the pair:
   KNO_3 or NaCl,
   NH_4Cl or $MgCO_3$,

MATERIALS

$FeCl_3$ or $FeCl_2$, K_2SO_4 or KCl, $LiNO_3$ or Na_2CO_3, $SrNO_3$ or Na_2SO_3, $(NH_4)_2CO_3$ or $ZnSO_4$, $BaCl_2$ or $NaNO_3$, Na_2SO_3 or Na_2CO_3, $Fe(NO_3)_2$ or $Zn(NO_3)_2$

PROCEDURE

Required Precautions

- Safety goggles and a lab apron must be worn at all times. All flame tests must be performed in the hood.

- Wear goggles, face shield, impermeable gloves, and a lab apron when you prepare the iron(III) chloride solution containing HCl. Work in a hood known to be in operating condition, with another person present nearby to call for help in case of an emergency. Be sure you are a 30 s walk from a safety shower and eyewash station known to be in operating condition.

Techniques to Demonstrate

Emphasize that each of these tests requires only a few crystals of sample. Caution students about not contaminating the unknown samples.

Call your teacher in the event of a spill. Spills should be cleaned up promptly, according to your teacher's directions.

When using a Bunsen burner, confine long hair and loose clothing. If your clothing catches on fire, WALK to the emergency lab shower, and use it to put out the fire. Do not heat glassware that is broken, chipped, or cracked. Use tongs or a hot mitt to handle heated glassware and other equipment because hot glassware does not look hot.

Never put broken glass into a regular waste container. Broken glass should be disposed of properly in the broken-glass waste container.

- 1.0 M HCl
- 1.0 M NaOH
- 0.1 M $KMnO_4$
- Bunsen burner and related equipment
- cobalt glass plates, 2
- filter paper, cut into short strips
- forceps
- Na_2CO_3
- Na_2SO_3
- NH_4Cl
- platinum wire, 5 cm length
- red litmus paper
- spot plate or small test tubes, 3
- sparker

1. Moisten a strip of filter paper with 0.1 M $KMnO_4$. Place a few crystals of sodium sulfite on a spot plate or in a small test tube, and add 2 drops of 1.0 M HCl. Immediately hold the strip of filter paper with tongs over the crystals, as shown in Figure A. This reaction is characteristic of the sulfite ion, SO_3^{2-}. Record your observations in the appropriate space in the Data Table.

FIGURE A

2. Place a few crystals of sodium carbonate on a spot plate or in a small test tube. Add 2 drops of 1.0 M HCl. With forceps, immediately hold a strip of filter paper moistened with 0.1 M $KMnO_4$ over the solution. This reaction is characteristic of the carbonate ion, CO_3^{2-}. Record your observations.

64 EXPERIMENT B10

ChemFile

EXPERIMENT B10 continued

Disposal

1. Provide labeled waste containers for the various solids. These can be saved for reuse.
2. Carry out the following procedures in the order stated.
   a. Combine all solutions containing Na_2SO_3, and, in a hood known to be operating properly, slowly add 1.0 M HCl while stirring until no more SO_2 is evolved.
   b. Combine all solutions containing NH_4^+, and, in a hood known to be operating properly, add an equal volume of 1.0 M NaOH solution. Then heat to boiling and continue to heat until all NH_3 is driven off.
   c. Combine all solutions containing Fe^{2+}, and add an excess of potassium hexacyanoferrate(III) to precipitate all Fe^{2+}.
   d. Combine all solutions containing Fe^{3+}, and add an excess of potassium hexacyanoferrate(II) to precipitate all Fe^{3+}.

3. Place a few crystals of ammonium chloride on a spot plate or in a small test tube. Add 2 drops of 1.0 M NaOH. Quickly hold a piece of moistened red litmus paper over the solution, as shown in Figure B. This reaction is characteristic of the ammonium ion, NH_4^+. Record your observations.

FIGURE B

4. You now have a number of simple tests to identify each of the unknowns listed in the data table. If a solution is needed for a test, dissolve a small amount of the compound in water. Consider all four ions in the two compounds for each unknown. For example, unknown 1 contains K^+, NO_3^-, Na^+, and Cl^- ions. You should be able to predict the results for each compound before you begin the test.

Cleanup and Disposal

5. Clean all apparatus and your lab station. Return equipment to its proper place. Dispose of chemicals and solutions in the containers designated by your teacher. Do not pour any chemicals down the drain or in the trash unless your teacher directs you to do so. Wash your hands thoroughly before you leave the lab and after all work is finished.

Data Table

	Test used	Observations	Identity
1. KNO_3 or NaCl			
2. NH_4Cl or $MgCO_3$			
3. $FeCl_3$ or $FeCl_2$			
4. K_2SO_4 or KCl			
5. $LiNO_3$ or Na_2CO_3			
6. $Sr(NO_3)_2$ or Na_2SO_3			
7. $(NH_4)_2CO_3$ or $ZnSO_4$			
8. $BaCl_2$ or $NaNO_3$			
9. Na_2SO_3 or Na_2CO_3			
10. $Fe(NO_3)_3$ or $Zn(NO_3)_2$			

EXPERIMENT B10 continued

e. Combine all solutions containing KMnO₄, and add 1 mL of 1.0 M HCl. Then, while stirring slowly, add 1.0 M Na₂S₂O₃ until the mixture is decolorized.

f. Combine all Zn²⁺ solutions. Add 1.0 M NaOH slowly with stirring until all zinc has precipitated as the hydroxide.

Observations, step 1

When acid is added to a sulfite, bubbles appear. These bubbles change the purple permanganate solution to a brown color.

Observations, step 2

When acid is added to a carbonate, bubbles appear. These bubbles do not affect the color of KMnO₄ solution.

Observations, step 3

The color of red litmus paper turns blue.

QUESTIONS

1. **Relating Ideas** Explain how to generate ammonia gas in the laboratory. Write and balance the equation for the reaction.

    Add sodium hydroxide to an ammonium salt, for example, ammonium chloride.

    $NH_4Cl(s) + NaOH(aq) \rightarrow NaCl(aq) + NH_3(g) + H_2O(l)$

2. **Relating Ideas** Give the formula and the name of a compound that gives a violet color to a flame and, when HCl is added, produces bubbles of gas that turn potassium permanganate brown.

    K_2SO_3, potassium sulfite

GENERAL CONCLUSIONS

1. **Analyzing Results** Which salt in the table will react with both HCl and NaOH? Write the equations for the reactions.

    Ammonium carbonate reacts with both HCl and NaOH.

    $(NH_4)_2CO_3(s) + 2HCl(aq) \rightarrow 2NH_4Cl(aq) + H_2O(l) + CO_2(g)$

    $(NH_4)_2CO_3(s) + 2NaOH(aq) \rightarrow 2NH_3(g) + 2H_2O(l) + Na_2CO_3(aq)$

Name _____

Date _____ Class _____

EXPERIMENT B11

Generating and Collecting O₂

OBJECTIVES

Recommended time:
20–25 minutes

- **Observe** the production of O$_2$.
- **Use** the technique of water displacement to collect O$_2$ gas.
- **Relate** chemical and physical properties to observations.
- **Apply** conclusions to larger settings.
- **Formulate** hypotheses.
- **Compare** reactions of O$_2$ gas in different situations.

INTRODUCTION

Materials (not in student's list)

- Micro O$_2$ generator vial with cap
- graduated pipet or thin-stemmed pipet

When oxygen reacts with another substance and large amounts of heat and light are generated, the reaction is called burning or combustion. In these reactions, oxygen is referred to as supporting combustion, because oxygen itself is not considered a combustible gas. What is the nature of combustion? How does combustion in an environment of 100% oxygen compare with combustion in air which is 20% oxygen? To answer these questions, you will generate and collect some oxygen gas and test it yourself to discover some of its chemical and physical properties.

SAFETY

Solution/Material Preparation

1. 3% hydrogen peroxide may be purchased from a local store, provided it is fresh.

2. To prepare 25 mL of 1.0 M KI solution, dissolve 0.42 g of KI in enough water to make 25 mL of solution. This will be far more than a class will use, so it can be kept for next year. Distribute the KI in dropper bottles or in cut-off thin-stem pipets. One bulb can hold about 80–120 drops, enough for the entire class.

Always wear safety goggles and a lab apron to protect your eyes and clothing. If you get a chemical in your eyes, immediately flush the chemical out at the eyewash station while calling to your teacher. Know the location of the emergency lab shower and eyewash station and the procedure for using them.

Do not touch any chemicals. If you get a chemical on your skin or clothing, wash the chemical off at the sink while calling to your teacher. Make sure you carefully read the labels and follow the precautions on all containers of chemicals that you use. If there are no precautions stated on the label, ask your teacher what precautions you should follow. Do not taste any chemicals or items used in the laboratory. Never return leftovers to their original containers; take only small amounts to avoid wasting supplies.

Call your teacher in the event of a spill. Spills should be cleaned up promptly, according to your teacher's directions.

When using a Bunsen burner, confine long hair and loose clothing. If your clothing catches on fire, WALK to the emergency lab shower, and use it to put out the fire.

EXPERIMENT B11 continued

⚠ A beaker of water should be kept nearby at all times, in case you need to quickly extinguish a burning wood splint.

MATERIALS

- 1.0 M KI solution
- 3% H_2O_2 solution (hydrogen peroxide)
- 100 mL beakers (or larger), 2
- Bunsen burner or candle
- forceps
- micro O_2 generator
- paper towels
- pegs, small, to fit holes in stoppers, 5
- petri dish, half
- rubber stoppers, one-holed, 5
- steel wool
- tap water
- small test tubes, 7
- wooden splints

PROCEDURE

3. To prepare the micro generator, cut the tapered tips off the graduated pipets to create the nozzles, or make nozzles by stretching thin-stemmed pipets, with a steady pull, to a much narrower diameter. If you do the latter, you can make two nozzles from each stem. Punch or drill holes (or melt them with a hot nail) in the center of the vial cap, and slip the tapered nozzles into the hole. If the nozzle segment is inserted until the flared-end section wedges into the hole, a relatively air-tight junction is formed.

The concentrations given for the H_2O_2 solution (3%) and the KI solution (1.0 M) are meant only as guidelines. You may wish to adjust them as needed to moderate the reaction rate. Having the solution at (or slightly above) room temperature will help speed the reaction.

Required Precautions

- Safety goggles and a lab apron must be worn at all times.

1. Label one beaker *water* and the other *spent solution*. Fill the appropriately labeled beaker with water. Be sure both beakers are nearby as you proceed.

2. Label the test tubes *1* through *7*. Fill test tubes *2*, *4*, *5*, *6*, and *7* completely with water. Insert the one-holed stoppers without creating any air bubbles in the test tubes. (Leave test tubes *1* and *3* empty and open in the test-tube rack—they will serve as controls.)

3. The micro O_2 generator is a plastic vial capped with a lid having a nozzle, as shown in Figure A.
Make sure your O_2 generator cap has a nozzle and that the nozzle is not plugged in any way before continuing the experiment.
Remove the cap, and carefully add enough 3% H_2O_2 to fill the vial to within 1 cm of the top. Then add 3 or 4 drops of KI solution. Replace the cap and set the O_2 generator in the petri dish. Record your observations.

4. To collect the gases, place test tube *2* mouth downward over the nozzle of the generator, as shown in Figure A.
As soon as the test tube is completely filled with oxygen, remove it from the nozzle and plug the stopper hole with a small peg to prevent the collected gas from escaping. Repeat this procedure for test tubes *4*, *5*, *6*, and *7*.

FIGURE A

EXPERIMENT B11 continued

- Read all safety precautions, and discuss them with your students.

- In case of a spill, use a dampened cloth or paper towel (or more than one towel if necessary) to mop up the spill. Then rinse the towel in running water at the sink, wring it out until it is only damp, and put it in the trash.

Techniques to Demonstrate

Demonstrate the use of the O_2 generator and the collection procedure.

Disposal

After the lab, solutions may be washed down the sink, provided your school drains are connected to a sanitary sewer system with a treatment plant.

5. If the reaction slows down so that it takes more than 1 min to collect a test tube of oxygen gas, uncap the vial, decant the liquid into the *spent solution* beaker, and replace it with fresh H_2O_2 solution and drops of KI solution. Then replace the cap and nozzle, and resume collecting gas.

   For steps 6–12, read the directions carefully, and predict what you think might happen. Write your predictions in the Data Table. Then follow the directions, and record your observations alongside your predictions.

6. Flaming wood splint and test tube *1* (control): Light the Bunsen burner or candle and use it to light one end of a wood splint.
   Keep the gas-generating vial away from the flame and wood splint!
   Hold test tube *1* (control test tube—filled with air) horizontally, and carefully insert the flaming wood splint. Record your observations in the Data Table.

7. Flaming wood splint and test tube *2* (oxygen): Relight the wood splint, remove the stopper from test tube *2*, hold the test tube horizontally, and carefully insert the burning wood splint. Record your observations.

8. Glowing wood splint and test tube *3* (control): Light the end of the wood splint again. Let it burn for 5 to 10 seconds, and then blow out the flame, but make sure a glowing ember persists on the tip of the wood splint. Hold test tube *3* (control test tube—filled with air) horizontally, and carefully insert the glowing wood splint. Record your observations.

9. Glowing wood splint and test tube *4* (oxygen): Repeat step **8** using test tube *4* and holding the test tube horizontally. Carefully unstopper the test tube before inserting the glowing wood splint. Record your observations.

10. Unstopper test tube *5* and hold it mouth upward for 2 minutes. Then try the glowing wood splint test as described in step **8**. Record your observations.

11. Unstopper test tube *6* and hold it mouth downward for 2 minutes. Then try the glowing wood splint test as described in step **8**. Record your observations.

12. Twist a few strands of steel wool into a single yarnlike string. Fold it in half, and hold the two loose ends together with forceps. Hold the folded bend in the Bunsen burner or candle flame for a few seconds, and then remove it and observe. Repeat this procedure, but this time quickly and carefully insert the glowing steel wool into test tube *7*. Record your observations.

13. If your teacher approves, refill the test tubes first with water and then with oxygen gas, as you did before, and recheck the observations you made.

Cleanup and Disposal

14. Clean all apparatus and your lab station. Return equipment to its proper place. Dispose of chemicals and solutions in the containers designated by your teacher. Do not pour any chemicals down the drain or in the trash unless your teacher directs you to do so. Make sure to shut off the gas valve completely before leaving the laboratory. Wash your hands thoroughly before you leave the lab and after all work is finished.

ChemFile

EXPERIMENT B11 continued

Data Table

Test-tube number	Predictions	Observations
1 flame, air		Flame burns normally, then dies out. Smoke fills the tube.
2 flame, O₂		Flame burns more intensely, then dies out.
3 ember, air		Ember glows for a while, then dies out.
4 ember, O₂		Glowing ember reignites, then dies out.
5 mouth up, ember		Ember glows for a while, then dies out.
6 mouth down, ember		Ember glows for a while, then dies out.
7 lit steel wool, O₂		Steel wool ember brightens, glows white-hot, and sparks fly.

QUESTIONS

1. **Analyzing Information** What evidence was there that a reaction was occurring inside the generator?

   The evidence for a chemical reaction is the appearance of bubbles.

2. **Relating Ideas** Write a balanced chemical equation for the O₂ generator reaction. (Hint: include KI in the equation, but only as a catalyst.)

   $$2H_2O_2(aq) \xrightarrow{KI} 2H_2O(l) + O_2(g)$$

3. **Relating Ideas** Write a word equation for the wood-splint tests using the following reactants and products: ash, wood, carbon dioxide, water vapor, and oxygen.

   wood + oxygen → ash + carbon dioxide + water vapor

GENERAL CONCLUSIONS

1. **Inferring Conclusions and Relating Ideas** What caused the burning wood splint to eventually be extinguished in test tube *1*? (Hint: refer to the word equation you wrote in answer to Question **3**.)

   The wood splint was extinguished because the oxygen in the air in the test tube was consumed and the combustion reaction could not continue.

EXPERIMENT B11 continued

2. Inferring Conclusions and Resolving Discrepancies Compare your results for test tubes *1* and *2*. Explain the differences you observed by considering which of the two reactants, wood or oxygen, is the limiting reactant for the reaction.

The wood splint burned less intensely in test tube 1 because the tube contained only the amount of oxygen present in air (20 percent), whereas test tube 2 contained 100 percent oxygen. The oxygen in test tube 1 was used up more quickly because there was less of it. Oxygen is the limiting reactant for both test tubes because unburned wood splint was left after the reaction was over.

3. Inferring Conclusions and Resolving Discrepancies Compare your results for test tubes *3* and *4*. Why didn't the ember in test tube *3* behave like the ember in test tube *4*? (Hint: Which reaction was quicker and more thorough? Why?)

The splint reignited in pure oxygen. Presumably the glowing ember was not hot enough to restart the reaction and rekindle the flame in air, but it was hot enough to rekindle the flame in an environment of 100 percent oxygen.

4. Inferring Conclusions and Resolving Discrepancies How does the "burning" of steel wool in air compare with the burning of steel wool in pure O_2 (test tube *7*)? Explain.

Steel wool burns much more readily in pure oxygen than in air. As discussed in General Conclusions **2**, this is reasonable because pure oxygen has five times as many oxygen molecules in a given volume available to react.

EXPERIMENT B11 continued

5. **Organizing Conclusions** Using your observations from this experiment, list as many physical properties of oxygen as you can. Explain what evidence you have for each property. (At least four properties were observable in this experiment.)

   Students' answers may vary. Four physical properties of oxygen

   that were observable in this experiment are

   • oxygen is colorless.

   • oxygen is odorless.

   • oxygen does not dissolve in water.

   • oxygen is about as dense as air.

6. **Predicting Outcomes** If you were to put a plain wood splint (with no ember or flame) into a tube full of oxygen, would it burst into flames? Would an ember form? Explain your answer. (Hint: what is required for a chemical reaction to occur in addition to having the reactant(s) present?)

   The wood splint would not burst into flames or form an ember by

   itself, because along with the reactants, a certain amount of

   energy or heat is needed to start the reaction. Once the reaction

   has begun, the energy released during the exothermic reaction

   serves as the activation energy for other molecules of wood and

   oxygen to react.

EXPERIMENT B12

Generating and Collecting H₂

OBJECTIVES

Recommended time:
25–30 minutes

- **Observe** the production of H_2.
- **Use** the technique of water displacement to collect H_2 gas.
- **Formulate** hypotheses for reactions of H_2 gas in different situations.

INTRODUCTION

Materials (not in student's list)

- Micro H₂ generator vial with cap graduated pipet or thin-stemmed pipet

Hydrogen gas has the lowest density of all the substances found on Earth. Once, hydrogen was used in blimps and dirigibles. Today, however, such aircraft use only helium, even though helium is less buoyant and considerably more difficult to obtain. Why was helium substituted for hydrogen? What was wrong with using hydrogen? To answer these questions, you will generate and collect small quantities of hydrogen gas and test the samples to find out for yourself some of the properties of hydrogen.

SAFETY

Solutions/Material Preparation

1. Wear goggles, a face shield, impermeable gloves, and a lab apron when you prepare the HCl. Work in a hood known to be in operating condition with another person present nearby to call for help in case of an emergency. Be sure you are within a 30 s walk of a safety shower and eyewash station known to be in operating condition.

2. To make 250 mL of 1.0 M HCl, observe the required safety precautions. Add 10.6 mL of concentrated HCl, while stirring, to enough distilled water to make 250 mL of solution.

Always wear safety goggles and a lab apron to protect your eyes and clothing. If you get a chemical in your eyes, immediately flush the chemical out at the eyewash station while calling to your teacher. Know the location of the emergency lab shower and eyewash station and the procedure for using them.

Do not touch any chemicals. If you get a chemical on your skin or clothing, wash the chemical off at the sink while calling to your teacher. Make sure you carefully read the labels and follow the precautions on all containers of chemicals that you use. If there are no precautions stated on the label, ask your teacher what precautions you should follow. Do not taste any chemicals or items used in the laboratory. Never return leftovers to their original containers; take only small amounts to avoid wasting supplies.

Call your teacher in the event of a spill. Spills should be cleaned up promptly, according to your teacher's directions.

When using a Bunsen burner, confine long hair and loose clothing. If your clothing catches on fire, WALK to the emergency lab shower, and use it to put out the fire. A beaker of water should be kept nearby at all times, in case you need to quickly extinguish a burning wood splint.

ChemFile

EXPERIMENT B12 **73**

EXPERIMENT B12 continued

MATERIALS

- 1.0 M HCl
- 100 mL or larger beakers, 2
- Bunsen burner or candle
- forceps or tongs (optional)
- micro H₂ generator
- pegs small enough to fit holes in stoppers, 5
- petri dish
- one-holed rubber stoppers, 5
- tap water
- small test tubes, 5
- wooden splints

PROCEDURE

3. To make a micro H₂ generator, cut the tapered tip off a graduated pipet to create a nozzle, or with a steady pull, stretch a thin-stemmed pipet to a much narrower diameter and cut it. (Each stem provides two nozzles.) Punch or drill a hole in the center of the vial cap (or melt it with a hot nail). Slip the tapered nozzle into the hole. If the nozzle segment is inserted until the flared-end section wedges into the hole, a relatively airtight junction is formed.

Break the zinc into pieces. Put about 0.15 g of zinc into each vial. The vials should be about one-sixth full.

A few drops of CuSO₄ solution will catalyze the Zn/HCl reaction. Also, the concentration of the HCl solution may be decreased to moderate the reaction rate.

4. To prepare 25 mL of 0.2 M copper(II) sulfate solution, dissolve 1.25 g of CuSO₄ • 5H₂O in enough distilled water to make 25 mL of solution.

Required Precautions
- Safety goggles and a lab apron must be worn at all times.
- Read all safety precautions, and discuss them with your students.

1. Label one beaker *water* and the other *spent solution*. Fill the appropriately labeled beaker with water. Be sure both beakers are nearby as you proceed.

2. Label the test tubes *1, 2, 3, 4,* and *5*. Fill test tubes 1, 3, 4, and 5 completely full of water. Fill test tube 2 only half way. Insert the one-holed stoppers into all five test tubes. Avoid creating air bubbles in the test tubes.

3. The micro H₂ generator is a plastic vial containing several pieces of zinc metal capped with a lid that has a nozzle.
   Make sure your H₂ generator cap has a nozzle and that the nozzle is not plugged before continuing with the experiment.
   Remove the cap, and carefully add enough 1.0 M HCl to fill the vial to within 1 cm of the top. Replace the cap, and set the H₂ generator in the petri dish. Observe the reaction, and record your observations in your lab notebook.

4. To collect the gas, place water-filled test tube 1 upside down over the nozzle of the generator, as shown in Figure A. As soon as the test tube is completely filled with hydrogen, remove it from the nozzle and plug the stopper hole with a small peg to prevent the collected gas from escaping. Repeat this procedure using test tubes 2, 3, 4, and 5.

FIGURE A

5. If the reaction slows down so that it takes more than 1 min to collect the tube of hydrogen gas, lift off the tube and uncap the vial. Decant the remaining liquid into a large beaker and replace it with fresh solution. Replace the cap and nozzle, and resume collecting gas.

EXPERIMENT B12 continued

- In case of a spill, use a dampened cloth or paper towel (or more than one towel if necessary) to mop up the spill. Then rinse the towel in running water at the sink, wring it out until it is only damp, and put it in the trash.

Techniques to Demonstrate

Show students how to manipulate the H_2 generator and how to collect the H_2 gas. Tell them whether you want them to hold the test tubes with their hand or use tongs when performing the pop tests. Demonstrate the technique for inserting the wooden splint.

Pre-lab Discussion

Encourage students to make predictions in the table before they begin the lab work.

Disposal

Combine all mixtures of zinc and acidic liquid. Filter to separate the zinc and wash the zinc at least three times with water, collecting the washings with the acidic filtrate. After the washed zinc is thoroughly dry, it may be saved for reuse.

For steps **6** through **10**, read the directions carefully and predict what might happen. Write your predictions in the Data Table. Then follow the directions, and record your observations alongside your predictions in the data table. **Because this investigation is carried out on a microscale level, the pop tests in steps 6 through 10, though potentially loud, are completely safe. The test tubes may be held with forceps or tongs or in your hand. Your teacher will tell you how you should hold the test tubes for the pop tests. Be sure to keep the gas-generating vial away from the flame and the wooden splint!**

6. Flaming wooden splint and test tube 1 (hydrogen): Light the Bunsen burner or candle, and use it to light one end of a wood splint. Remove the stopper from test tube 1, and carefully insert the burning splint into the mouth of the test tube. Record your observations in the Data Table.

7. Flaming wooden splint and test tube 2 (half hydrogen): Repeat the pop test described in step 6 with test tube 2, which contains only half as much hydrogen as test tube 1 did. Record your observations in the Data Table.

8. Glowing wooden splint and test tube 3 (hydrogen): Repeat the pop test with test tube 3, but this time blow out the flame, making sure a glowing ember persists on the tip of the wooden splint as you insert it into the test tube. Record your observations.

9. Unstopper test tube 4 and hold it mouth upward for 30 s; then try the pop test. Record your observations.

10. Unstopper test tube 5 and hold it mouth downward for 30 s; then try the pop test. Record your observations.

11. If your teacher approves, refill the test tubes first with water and then with hydrogen gas and recheck any of the observations you made. Record all observations.

Cleanup and Disposal

12. Clean all apparatus and your lab station. Return equipment to its proper place. Dispose of chemicals and solutions in the containers designated by your teacher. Do not pour any chemicals down the drain or in the trash unless your teacher directs you to do so. Shut off the gas valve completely before leaving the laboratory. Wash your hands thoroughly before you leave the lab.

Data Table

Test tube number	Predictions	Observations
1 flame, pure H_2		small pop (can be repeated)
2 flame, 50% H_2		louder, sharper pop (only one pop possible)
3 ember, pure H_2		glowing ember extinguished, no pop
4 mouth up, flame		no pop
5 mouth down, flame		loud pop comparable to test tube 2 (probably only one pop possible)

EXPERIMENT B12 continued

QUESTIONS

Neutralize the combined filtrate and water washes with 0.1 M NaOH. Filter; the filtrate may be poured down the drain. After the Zn(OH)$_2$ precipitate has dried, it may be wrapped in newspaper and discarded in the trash.

1. **Analyzing Information** What evidence was there that a reaction was occurring inside the generator?

   The evidence for a chemical reaction includes the appearance of bubbles and the slow disappearance of zinc metal.

2. **Relating Ideas** Write a balanced chemical equation for the H$_2$ generator reaction.

   2HCl(aq) + Zn(s) → ZnCl$_2$(aq) + H$_2$(g)

3. **Relating Ideas** Write a balanced chemical equation for the reaction taking place during the pop test. (Hint: Water vapor is the product of the reaction.)

   O$_2$(g) + 2H$_2$(g) → 2H$_2$O(g). Students may also include heat or enthalpy evolved as a part of the equation.

GENERAL CONCLUSIONS

1. **Inferring Conclusions and Resolving Discrepancies** Compare your results for test tubes 1 and 2. Test tube 2 had only half as much hydrogen; was the pop test only half as loud? Explain your answer, referring to the balanced chemical equation for the pop test.

   The poptest for test tube 2 was louder than that for test tube 1. The balanced equation shows that the pop test involves both H$_2$ and O$_2$. In the first tube, the pure hydrogen reacts with oxygen in the air near the opening of the tube. The second tube is filled with a mixture of hydrogen and air, so oxygen is available throughout the tube for a rapid and thorough reaction.

2. **Inferring Conclusions and Resolving Discrepancies** Consider your results for test tube 3. Did the hydrogen ignite? Can you explain why you observed what you did?

   In test tube 3, the hydrogen did not ignite. Instead, it extinguished the ember. One of the chemical properties of hydrogen is that it will combine with oxygen if it is heated above 800°C or ignited by a spark. Apparently, the ember did not provide enough energy to initiate the pop test reaction.

EXPERIMENT B12 continued

3. Inferring Conclusions and Resolving Discrepancies Is hydrogen gas more dense or less dense than air? (Hint: compare your results for test tubes 4 and 5. What was different about how they were treated, and how did this affect their pop test results?)

Hydrogen gas is less dense than air. When test tube 4 was held mouth up, much of the hydrogen gas floated away. When test tube 5 was held mouth down, much of the hydrogen gas remained in the tube. This suggests that air is denser than hydrogen.

4. Organizing Conclusions Using your observations of the hydrogen gas in each trial, the generation reaction, and the poptests, list as many chemical properties of hydrogen as you can. Explain what evidence you have for each one. (At least four properties were observable in the experiment.)

Students' answers may vary. Four chemical properties of hydrogen that were observable in this experiment are

- hydrogen is replaced in acids by zinc.
- hydrogen reacts with oxygen in the air if ignited by a flame.
- hydrogen will not ignite and react with oxygen in the air if a red-hot ember is used.
- hydrogen alone cannot support combustion of an ember.

5. Organizing Conclusions Using your observations from this experiment, list as many physical properties of hydrogen as your can. Explain what evidence you have for each one. (At least four properties were observable in this experiment.)

Students' answers may vary. Four physical properties of hydrogen that were observable in this experiment are

- hydrogen is colorless.
- hydrogen is odorless.
- hydrogen does not dissolve in water.
- Hydrogen is less dense than air.

EXPERIMENT B12 continued

6. **Predicting Outcomes and Designing Experiments** Air is about 20% oxygen. Use this information to predict what ratio of air to hydrogen will provide the loudest pop test. Design an experiment that will test a range of proportions, including the one that you predicted would be the loudest. If your teacher approves your suggestions, test your predictions.

Students' answers will vary, but they should suggest that a proportion of about 5 parts air to 2 parts hydrogen should be the most effective. In this proportion, the oxygen present in the air should react completely with the hydrogen, maximizing the production of the products: water vapor, heat, and sound.

Name _____

Date _____ Class _____

EXPERIMENT B13

HOLT ChemFile LAB PROGRAM

Testing Reaction Combinations of H_2 and O_2

OBJECTIVES

Recommended time: 50 min, including rocket launching and other suggested activities

- **Observe** the production of H_2 and O_2.
- **Use** the technique of water displacement to collect H_2 and O_2 gas.
- **Relate** chemical concepts such as stoichiometry, limiting reagent, activation energy, and thermodynamics to observations of chemical reactions.
- **Infer** a conclusion from experimental data and **evaluate** methods.

INTRODUCTION

Materials (not in student's list)

- micro H_2 generator graduated pipet or thin-stemmed pipet vial with cap
- micro O_2 generator graduated pipet or thin-stemmed pipet vial with cap

Hydrogen, H_2, and oxygen, O_2, are two gases that react with each other in a very quick, exothermic (heat-producing) manner. The explosiveness of this reaction is greatest when hydrogen and oxygen are present in just the proper proportion. The reaction is used to power three of the rocket engines that carry the space shuttle into orbit.

In this experiment, you will generate hydrogen and oxygen and collect them using the "water displacement" method. You will test their explosive nature, first separately and then in reactions of different proportions of the two gases. After finding the most powerful reaction combination, you will use this reaction to launch a rocket across the room.

SAFETY

- piezoelectric sparking element made from piezoelectric butane charcoal/fireplace lighters (such as Aim 'n Flame)
- 10 cm of 24 AWG speaker hook-up wire (solid not stranded: Radio Shack #278-1509)
- film canister lid
- electrician's tape
- hot glue gun and glue
- scissors
- wire strippers
- 1/8 inch drill or 6-D common nail
- tongs or forceps (optional)

Always wear safety goggles and a lab apron to protect your eyes and clothing. If you get a chemical in your eyes, immediately flush the chemical out at the eyewash station while calling to your teacher. Know the location of the emergency lab shower and eyewash station and the procedure for using them.

Do not touch any chemicals. If you get a chemical on your skin or clothing, wash the chemical off at the sink while calling to your teacher. Make sure you carefully read the labels and follow the precautions on all containers of chemicals that you use. If there are no precautions stated on the label, ask your teacher what precautions you should follow. Do not taste any chemicals or items used in the laboratory. Never return leftovers to their original containers; take only small amounts to avoid wasting supplies.

Call your teacher in the event of a spill. Spills should be cleaned up promptly, according to your teacher's directions.

ChemFile   EXPERIMENT B13   **79**

EXPERIMENT B13 continued

MATERIALS

For General Conclusions item 4
- refrigerator for chilling gases

- 1.0 M HCl solution
- 1.0 M KI solution
- 3% H$_2$O$_2$ solution
- 100 mL beakers (or larger), 2
- collection bulb, calibrated
- micro H$_2$ generator
- micro O$_2$ generator
- petri dish, half
- Piezoelectric sparking element
- mossy zinc
- tongs or forceps
- tap water

PROCEDURE

Solution/Material Preparation

1. To make 250 mL of 1.0 M HCl, observe the required safety precautions. Add 10.6 mL of concentrated HCl, with stirring, to enough distilled water to make 250 mL of solution.

2. 3% hydrogen peroxide may be store-bought provided it is fresh.

3. To prepare 25 mL of 1.0 M KI solution, dissolve 0.42 g of KI in enough water to make 25 mL of solution. This will be far more than a class will use, so it can be kept for next year. Distribute the KI in dropper bottles, or in cut-off thin-stem pipets. One bulb can hold about 80–120 drops, which may be enough for the entire class.

4. Make collection bulbs by cutting off the bulbs of graduated or thin-stemmed pipets. To calibrate the bulbs, fill one bulb completely with water. (If you have trouble filling the bulb completely, or students do later, consider using a thin-stemmed pipet to deliver the water into the bulb instead of having the water squeezed up from the petri dish.) Once the bulb is filled, squeeze all the water out into a 10 mL graduated cylinder, and record the volume. Empty the graduated cylinder and refill the bulb with water. Squeeze out, into the graduated cylinder, one-sixth of the volume

1. Label the two beakers *water* and *spent solution*. Be sure both beakers are nearby as you proceed. Fill the beaker labeled *water* full of water and the petri dish three-fourths full of water.

2. The micro H$_2$ generator is a plastic vial containing several pieces of zinc metal, capped with a lid and nozzle.
   Before continuing with the experiment make sure both generator caps have a nozzle and that the nozzle is not plugged in any way. Remove the cap of the H$_2$ generator, and carefully add enough 1.0 M HCl to fill the vial to within 1 cm of the top. Replace the cap and set the H$_2$ generator in the petri dish. Observe the reaction, and record your observations.

Observations:

_____

_____

_____

3. The micro O$_2$ generator is a plastic vial capped with a lid having a nozzle, as shown in Figure A. Remove the cap, and carefully add enough 3% H$_2$O$_2$ to fill the vial to within 1 cm of the top. Then add 3 or 4 drops of KI solution to serve as a catalyst. Replace the cap and set the O$_2$ generator in the petri dish beside the H$_2$ generator. Record your observations about the reaction in your lab notebook.

FIGURE A

4. Fill the collection bulb completely with water from the petri dish. The best way to do this is to squeeze the bulb tightly, invert it into the petri dish, and draw up as much water as possible. Then hold the bulb mouth upward and squeeze out the remaining air. Without letting go of the bulb, invert it again into the petri dish and draw up the remainder of the water needed to fill the bulb.

EXPERIMENT B13 continued

of water you recorded earlier. Mark the water level on the outside surface of the bulb when held mouth-downward. Squeeze out a second sixth of the volume, and mark the level again. Repeat for the remainder of the water. When one bulb has been calibrated, rest a wood splint across it, with the splint's end even with the end of the bulb. Mark off the splint at the same places that the bulb is marked. Use the splint as a template for calibrating the rest of the bulbs. Once calibrated, the bulbs can be reused from class to class and year to year.

5. To make a micro H_2 or O_2 generator, cut the tapered tip off a graduated pipet to create a nozzle, or, with a steady pull, stretch a thin-stemmed pipet to a much narrower diameter and cut it as shown in Figure 1 on page 86. (Each stem provides two nozzles.) Punch or drill a hole (or melt it with a hot nail) in the center of the vial cap. Slip the tapered nozzle into the hole. If the nozzle segment is inserted until the flared-end section wedges into the hole, a relatively airtight junction is formed.

5. To collect hydrogen gas, place the water-filled collection bulb from step **4** mouth downward over the nozzle of the H_2 generator, as shown in Figure A. As soon as the bulb is completely filled with hydrogen, remove it from the nozzle and place a finger over the mouth of the bulb to prevent the collected gas from escaping.

6. Test the bulb of hydrogen gas by moving your finger aside and inserting the wire tip of the piezoelectric sparker into the bulb. **Keep the piezoelectric sparker away from the gas-generating vials. Hold the bulb securely. (Your teacher will tell you whether you should hold the bulb with your hand, with tongs, or with forceps.)** Pull the trigger of the piezoelectric sparker and observe. Record the loudness of the reaction (on a scale of 1 to 10) in the Data Table.

7. To collect oxygen gas, refill the bulb with water, place the bulb mouth downward over the nozzle of the O_2 generator, and repeat steps **5** and **6**, pop-testing the bulb with the piezoelectric sparker. Record the loudness of the reaction (on a scale of 1 to 10) in the Data Table.

8. Refill the bulb with water, and begin collecting another bulb of O_2, but when the bulb is about one-sixth full of O_2 (with five-sixths water), move the bulb from the O_2 generator to the H_2 generator, and continue collecting. When the bulb is filled with gas, you will have a combination that is 1 part oxygen and 5 parts hydrogen. Remove the bulb from the nozzle, test it as in step **6**, and record the relative loudness of the reaction in the Data Table.

9. Repeat step **8**, making the switch from the O_2 to the H_2 generator at different times so that you can test various proportions (2:4, 3:3, 4:2, and 5:1) to complete the data table. Determine the optimum (most explosive) combination.

10. If either generator reaction slows down so that it takes more than 1 minute to fill the bulb with gas, lift off the bulb, uncap the vial, and decant the remaining liquid into the *spent solution* beaker. Refill the vial with fresh solution(s) from the appropriate bottle(s), replace the cap, and resume collecting gas.

11. Rocket Launch (check with your teacher before performing this step): Collect the optimum (loudest) mixture one more time, and instead of holding onto the bulb, aim it outward at a target chosen by your teacher. **Make sure nobody is in the line of fire.** Pull the trigger and launch your micro-rocket. Can you think of ways to make your rocket go farther? Try them!

Cleanup and Disposal

12. Clean all apparatus and your lab station. Return equipment to its proper place. Dispose of chemicals and solutions in the containers designated by your teacher. Do not pour any chemicals down the drain or in the trash unless your teacher directs you to do so. Wash your hands thoroughly before you leave the lab and after all work is finished.

Data Table

Parts H_2	6	5	4	3	2	1	0
Parts O_2	0	1	2	3	4	5	6
Relative loudness (rated from 0–10)	0	7	10	8	5	2	0

EXPERIMENT B13 continued

QUESTIONS

For the H₂ generator, break the zinc into pieces. Put about 0.15 g of zinc into each vial. The vials should be about one-sixth full. A few drops of CuSO₄ solution will catalyze the Zn/HCl reaction. The concentration of the HCl solution may be decreased to moderate the reaction rate.

For the O₂ generator, the concentrations given for the H₂O₂ solution (3%) and the KI solution (1.0 M) are meant only as guidelines. You may wish to adjust them as needed to moderate the reaction rate. Having the solution at (or slightly above) room temperature will also help speed the reaction.

6. To prepare 25 mL of 0.2 M copper(II) sulfate solution, dissolve 1.25 g of CuSO₄ • 5H₂O in enough distilled water to make 25 mL of solution.

7. Construct a piezoelectric sparker from a charcoal/fireplace lighter found in many stores. Once wired as described in the following instructions, it can be used in place of a Tesla coil for a variety of chemical demonstrations. If the wire and tape are removed, the sparker can be used as a Bunsen burner lighter, or for any demonstration requiring a quick and safe light. The materials are listed for making one piezoelectric sparker. Four or five sparkers are sufficient for an entire class. See the diagrams on page 86.

Using scissors, split the end of a 10 cm length of speaker wire down the center for a distance of 3.5 cm. Strip the last 2.3 cm of insulation from both sides of the fork, as shown in Figure 2.

1. Relating Ideas Write a balanced equation for the reaction
   a. taking place inside the O₂ generator.
   b. taking place inside the H₂ generator.

a. $2H_2O_2(aq) \xrightarrow{KI} 2H_2O(l) + O_2(g)$

b. $2HCl(aq) + Zn(s) \rightarrow ZnCl_2(aq) + H_2(g)$

2. Interpreting Ideas Which do you think will be used up first: the KI solution in the O₂ generator or the Zn in the H₂ generator? Why?

The Zn will need to be replaced first, because it is being consumed as a reactant. The KI is acting as a catalyst and theoretically should never need to be replaced.

3. Analyzing Methods What laboratory techniques did you use to help ensure that the gas collecting bulb contained only the desired gas?

Students' answers should include the following. The generators are filled so full that there is little room for air, so most of the gas that comes from the nozzle is pure oxygen or hydrogen. The collection bulbs are filled with water, so the collected gases are as pure as when they come from the nozzle. When the bulb is removed, it is promptly covered with a finger to prevent air from diffusing into the bulb.

4. Analyzing Results Make a bar graph of the relative loudness produced by your pop tests of the mixtures, with the loudness from 0 through 10 on the y-axis and trial 1 through 7 on the x-axis.

| Parts O₂ | 6 | 5 | 4 | 3 | 2 | 1 | 0 |
| Parts H₂ | 0 | 1 | 2 | 3 | 4 | 5 | 6 |

EXPERIMENT B13 continued

GENERAL CONCLUSIONS

Drill a 1/8 in. hole (or melt a hole with a hot 6D common nail) through the center of the film canister lid. Slide the wire through, and secure the lid in place about 4 cm from the unforked end of the wire with hot glue or other waterproof adhesive. The film canister lid serves as both a launch support pad and a splash guard.

Slide one of the two stripped wire ends down into the small butane nozzle of the charcoal lighter, all the way down, so that the insulation is against the nozzle. (Rotate the wire as you insert it.) Lay the other stripped end along the metal barrel of the lighter and secure it in place with 3–4 wrappings of electrician's tape.

To use the sparker, put the bulb over the wire, aim, and pull the trigger. Rather than arcing at the butane nozzle, the device will arc at the tip of the wire (the path of least resistance).

Required Precautions

- Safety goggles and a lab apron must be worn at all times.
- Read all safety precautions, and discuss them with your students.
- In case of a spill, use a dampened cloth or paper towel (or more than one towel if necessary) to mop up the spill. Then rinse the towel in running water at the sink, wring it out until it is only damp, and put it in the trash.

1. **Inferring Conclusions and Relating Ideas** From your observations, state the relative combustibility of the following:
   a. pure O_2
   b. pure H_2

   a. Oxygen does not pop or explode when tested. Oxygen supports combustion of other materials but is not itself combustible.

   b. Hydrogen is combustible, meaning that it reacts readily with oxygen. Pure hydrogen does not explode, because there is no oxygen present for it to react with.

2. **Resolving Discrepancies and Inferring Relationships** Were there any reaction combinations that produced no reaction at all? Explain what happened.

   Some mixtures contain so little of one of the reactants that the likelihood of an effective collision between an O_2 molecule and a H_2 molecule is slight. Thus, the reaction doesn't propagate throughout the mixture in a chain reaction.

3. **Inferring Conclusions and Inferring Relationships** What proportion of oxygen to hydrogen produced the most explosive reaction? Explain why that combination was the most explosive by referring to the balanced chemical equation for the reaction and the concepts of limiting reagents and maximum yields.

   The most explosive mixture is 2 parts O_2 to 4 parts H_2. This is the same as the 1:2 mole ratio in the balanced equation for the pop test. It is the most explosive combination because it produces the maximum yield of the product, water, and the maximum amount of released energy. In the 2:4 mixture, all the reactants are consumed, whereas in all other mixtures, one of the reactants is a limiting reactant.

ChemFile   EXPERIMENT B13   **83**

EXPERIMENT B13 continued

- If you divide the reagents up into several bottles of each and set up "reagent stations" in different parts of the lab, students will not have far to walk after obtaining chemicals, and the likelihood of spills and accidents will be decreased.

- Wear goggles, face shield, impermeable gloves, and a lab apron when you prepare the HCl. Work in a hood known to be in operating condition, with another person present nearby to call for help in case of an emergency. Be sure you are within 30 s walking distance of a safety shower and eyewash station known to be in operating condition.

Techniques to Demonstrate

If students are unfamiliar with the O_2 and H_2 generators, demonstrate how they are used and how to replenish them, if necessary. Show students how to perform the pop test using the piezoelectric sparker.

If piezoelectric sparking elements are not available, the gases can be pop-tested with a Bunsen burner, but the danger of unexpected explosions and burning bulbs is greatly increased. Therefore, if students are to use Bunsen burners, demonstrate this procedure: Light the burner and adjust it to a cool flame. When the bulb is filled, remove it from the nozzle, and hold it securely in a horizontal position with its mouth 1 cm from the midsection of the flame. Gently squeeze a very small portion of the contents of the bulb into the flame. If burners are used, it would be best to omit the rocket portion of the experiment.

4. Predicting Outcomes Would your results be different if you performed the same experiment with the gases chilled? If your teacher approves, test your prediction. Explain your results.

According to the kinetic theory of matter, the chilled gas molecules would have less kinetic energy. The molecules would collide less frequently and require more additional energy to become activated. If students test this, however, the difference may not be noticeable because the spark or flame is sufficient to provide the increased activation energy and initiate the chain reaction.

5. Resolving Discrepancies and Relating Ideas Why don't O_2 and H_2 react as soon as they mix in the collection bulb? The individual O_2 and H_2 molecules are certainly colliding with one another. What role does the spark or flame play?

O_2 and H_2 molecules collide as they mix in the bulb but do not have the extra energy needed to react (activation energy). The spark supplies that energy in the form of electricity or heat.

6. Evaluating Information Share your data with the rest of the class. Calculate the average relative loudness values for each mixture for the entire class, and make a bar graph of the results.

Students answers will vary. The class average for relative loudness should reach a maximum at 2 parts O_2 to 4 parts H_2. The average should reach minimums for pure O_2 and pure H_2.

EXPERIMENT B13 continued

*Show students how to fill the collection bulb from the petri dish as described in step **5**. If this proves difficult, you could consider using thin-stemmed pipets to deliver the water into the collection bulb.*

Pre-lab Discussion

This experiment involves a number of concepts that could be discussed: activation energy, exothermic reactions, stoichiometry, rocketry and propulsion, and why the ratio of the volumes of the two gases in the optimum combination turns out to be the same as the mole ratio in the chemical equation. The last concept provides an opportunity to preview Avogadro's principle and the concept of the molar volume of gases.

Disposal

Oxygen: After the lab, solutions may be washed down the sink, provided your school drains are connected to a sanitary sewer system with a treatment plant.

Hydrogen: Combine all mixtures of zinc and acidic liquid. Filter to separate the zinc and wash the zinc at least three times with water, collecting the washings with the acidic filtrate. After the washed zinc is thoroughly dry, it may be saved for reuse.

Neutralize the combined filtrate and water washes with 0.1 M NaOH. Filter; the filtrate may be poured down the drain. When the $Zn(OH)_2$ precipitate is dry, it may be wrapped in newspaper and discarded in the trash.

---

7. Analyzing Methods and Inferring Conclusions What methods did you attempt for making your rocket fly farther? Which methods worked best? Explain why they worked.

Students' answers will vary. Along with the ratio of the gases, the angle of the launch, the mass of the rocket, and the shape of the rocket also play a part in determining the distance the rocket will travel. Although a 45° angle is best under frictionless conditions, an angle of less than 45° is more effective with air resistance. If the rocket has a larger mass, it will be less subject to air resistance but will have a greater inertia to overcome. Streamlining the rocket (with tail fins and nose cone) can also increase the flight. If some water is left in the bulb, it will greatly increase the flight, because it gives the expanding gases a propellant to push against.

8. Applying Models The space shuttle carries 680 000. kg of fuel for its main engines. These engines use liquid hydrogen and liquid oxygen. Based on your results in this experiment, what is the most effective way to divide this mass? How much of it should be liquid oxygen, and how much liquid hydrogen?

The ratio found in this experiment, 1:2 (2:4), the **mole ratio**, is the most efficient for the space shuttle's fuel. Students can calculate the mass ratio as 8:1. When applied to the 680 000 kg of fuel, 603 900 kg should be oxygen, and 76 100 kg should be hydrogen.

This is equivalent to 11.19% H_2, 88.81% O_2

$$680\ 00\text{ kg total} \times \frac{11.19\text{ g }H_2}{100\text{ g total}} = 76\ 100\text{ kg }H_2$$

$$680\ 00\text{ kg total} \times \frac{88.81\text{ g }O_2}{100\text{ g total}} = 603\ 900\text{ kg }O_2$$

EXPERIMENT B13 continued

Graduated pipet

Collection bulb | Cut | Nozzle | Cut

Thin-stemmed pipet

Collection bulb | Cut | Stretch
Cut here for nozzle

FIGURE 1

a. Speaker wire — 3.5 cm

b. 2.3 cm

c. Lid to film container

d. 4 cm — Butane nozzle of charcoal fireplace lighter

e. Spark gap — Glue

f. Collection bulb — Tape

FIGURE 2

86 EXPERIMENT B13

Name _____

Date _____ Class _____

HOLT ChemFile
LAB PROGRAM

EXPERIMENT B14

Reacting Ionic Species in Aqueous Solution

OBJECTIVES

Recommended time: 20 min

- **Organize** experimental variables to study them one at a time.
- **Recognize** a precipitation reaction.
- **Formulate** net ionic equations for precipitation reactions from experimental data.

INTRODUCTION

Materials per lab group

- 0.1 M BaCl$_2$ solution, 1–2 drops
- 0.1 M Ba(NO$_3$)$_2$ solution, 1–2 drops
- 0.1 M Na$_2$C$_2$O$_4$ solution, 1–2 drops
- 0.1 M NaCl solution, 1–2 drops
- 0.1 M NaNO$_3$ solution, 1–2 drops
- 0.1 M Na$_2$SO$_4$ solution, 1–2 drops
- plastic sheet

When ionic solids dissolve in water, they dissociate into positive *cations* and negative *anions*. If two solutions containing dissolved ionic solids are mixed together, new combinations of cations and anions are possible. Sometimes a new combination of ions is not soluble in water, and a precipitate (solid) forms. For example, silver nitrate in solution dissociates into Ag$^+$ and NO$_3^-$ ions, and KCl in solution breaks up into K$^+$ and Cl$^-$ ions. If the two solutions are mixed, the positive Ag$^+$ ions combine with the negative Cl$^-$ ions to form an insoluble precipitate of AgCl. The *complete ionic* equation for the reaction is the following.

$$Ag^+(aq) + NO_3^-(aq) + K^+(aq) + Cl^-(aq) \rightarrow AgCl(s) + K^+(aq) + NO_3^-(aq)$$

Because the K$^+$ ion and NO$_3^-$ ion do not take part in the reaction, they are called spectator ions. Subtraction of the spectator ions from both sides of the equation produces the *net ionic* equation for the reaction.

$$Ag^+(aq) + Cl^-(aq) \rightarrow AgCl(s)$$

In this experiment, you will mix six different ionic solutions, two at a time, and observe which combinations form precipitates. You will also determine which pairs of ions form precipitates and write net ionic equations for the reactions.

SAFETY

Solution/Material Preparation

1. To prepare 25 mL of 0.1 M BaCl$_2$, add 0.6 g of BaCl$_2$ • 2H$_2$O to enough distilled water to make 25 mL of solution.

2. To prepare 25 mL of 0.1 M Ba(NO$_3$)$_2$, add 0.675 g of Ba(NO$_3$)$_2$ to enough distilled water to make 25 mL of solution.

For this experiment, wear safety goggles, gloves, and a lab apron to protect your eyes, hands, and clothing. If you get a chemical in your eyes, immediately flush the chemical out at the eyewash station while calling to your teacher. Know the location of the emergency lab shower and eyewash station and the procedure for using them.

Do not touch any chemicals. If you get a chemical on your skin or clothing, wash the chemical off at the sink while calling to your teacher. Make sure you carefully read the labels and follow the precautions on all containers of chemicals that you use. If there are no precautions stated on the label, ask your teacher what precautions you should follow. Never return leftovers to their original containers; take only small amounts to avoid wasting supplies.

Call your teacher in the event of a spill. Spills should be cleaned up promptly, according to your teacher's directions.

ChemFile   EXPERIMENT B14   **87**

EXPERIMENT B14 continued

MATERIALS

- Plastic sheet
- 0.1 M BaCl$_2$
- 0.1 M Ba(NO$_3$)$_2$
- 0.1 M Na$_2$C$_2$O$_4$
- 0.1 M NaCl
- 0.1 M NaNO$_3$
- 0.1 M Na$_2$SO$_4$

PROCEDURE

3. To prepare 50 mL of 0.1 M Na$_2$C$_2$O$_4$, dissolve 0.65 g of Na$_2$C$_2$O$_4$ in enough distilled water to make 50 mL of solution.

4. To prepare 100 mL of 0.1 M NaCl, dissolve 0.6 g of NaCl in enough distilled water to make 100 mL of solution.

5. To prepare 50 mL of 0.1 M NaNO$_3$, dissolve 0.45 g of NaNO$_3$ in enough distilled water to make 50 mL of solution.

6. To prepare 100 mL of 0.1 M Na$_2$SO$_4$, dissolve 2.8 g of Na$_2$SO$_4$ in enough distilled water to make 100 mL of solution.

7. Distribute the solutions in 30 mL dropper bottles.

8. A clear acetate sheet used for an overhead-projector transparency, plexiglass, or plastic wrap may be used for the plastic sheet.

Required Precautions

- Safety goggles and a lab apron must be worn at all times.

QUESTIONS

- Read all safety precautions, and discuss them with your students.

- In case of a spill, use a dampened cloth or paper towel (or more than one towel if necessary) to mop up the spill. Then rinse the towel in running water at the sink, wring it out until it is only damp, and put it in the trash.

1. Decide on the order in which you will perform the tests before you begin. Record your observations in the Data Table after you have performed each test. For each possible combination, indicate *PPT* if a precipitate forms and *clear* if no precipitate forms.

2. Place a drop or two of one of the solutions on the plastic sheet. Add a drop or two of the second solution. Do not allow the tip of the dropper to touch the drops on the plastic sheet. Record your result.

3. Choose a clean spot on the plastic and repeat step **2** for another pair of solutions. Repeat for all possible combinations indicated in your data table, and record your results.

Cleanup and Disposal

4. Clean all apparatus and your lab station. Return equipment to its proper place. Dispose of chemicals and solutions in the containers designated by your teacher. Do not pour any chemicals down the drain or in the trash unless your teacher directs you to do so. Wash your hands thoroughly before you leave the lab and after all work is finished.

Data Table

	Na$^+$, Cl$^-$	Na$^+$, NO$_3^-$	Na$^+$, C$_2$O$_4^{2-}$	Na$^+$, SO$_4^{2-}$	Ba^{2+}, Cl$^-$
Ba^{2+}, NO$_3^-$	clear	clear	PPT	PPT	clear
Ba^{2+}, Cl$^-$	clear	clear	PPT	PPT	
Na$^+$, SO$_4^{2-}$	clear	clear	clear		
Na$^+$, C$_2$O$_4^{2-}$	clear	clear			
Na$^+$, NO$_3^-$	clear				

1. **Analyzing Information** When a precipitate forms, you can assume that ions have combined to form an insoluble compound. For example, if CaCl$_2$ is mixed with Na$_2$SO$_4$, a precipitate forms. The possible combinations that could have produced the precipitate are CaSO$_4$ and NaCl. Write the possible formulas for each precipitate that you observed.

NaNO$_3$ and BaC$_2$O$_4$

NaNO$_3$ and BaSO$_4$

NaCl and BaC$_2$O$_4$

NaCl and BaSO$_4$

EXPERIMENT B14 continued

Techniques to Demonstrate

If necessary, show students how to avoid contaminating the solutions by holding the dropper 1 to 2 cm above the plastic sheet when adding one solution to another.

Pre-lab Discussion

Little or no discussion is needed for this discovery experiment.

Disposal

Combine all solutions and precipitates. Dilute tenfold with water and pour down the drain.

2. **Analyzing Information** For each pair of possible precipitates listed in item **1**, eliminate one of them using your data on solubility. For example, NaCl, one of the two possible combinations for item **1**, can be eliminated because it is also one of the original clear solutions. In a similar manner, list all the combinations that can be eliminated as possible precipitates.

NaCl and NaNO₃ were both original clear solutions, so they can

be eliminated.

_____

3. **Organizing Conclusions** Write a balanced *complete ionic* equation for each precipitate formed. Use the *complete* ionic equation in the Introduction as a guide.

$Ba^{2+}(aq) + 2NO_3^-(aq) + 2Na^+(aq) + C_2O_4^{2-}(aq) \rightarrow$

$BaC_2O_4(s) + 2Na^+(aq) + 2NO_3^-(aq)$

$Ba^{2+}(aq) + 2NO_3^-(aq) + 2Na^+(aq) + SO_4^{2-}(aq) \rightarrow$

$BaSO_4(s) + 2Na^+(aq) + 2NO_3^-(aq)$

$Ba^{2+}(aq) + 2Cl^-(aq) + 2Na^+(aq) + C_2O_4^{2-}(aq) \rightarrow$

$BaC_2O_4(s) + 2Na^+(aq) + 2Cl^-(aq)$

$Ba^{2+}(aq) + 2Cl^-(aq) + 2Na^+(aq) + SO_4^{2-}(aq) \rightarrow$

$BaSO_4(s) + 2Na^+(aq) + 2Cl^-(aq)$

4. **Organizing Conclusions** Write a *net ionic* equation for each reaction in which a precipitate is formed. Use the net ionic equation in the Introduction as a guide.

$Ba^{2+}(aq) + C_2O_4^{2-}(aq) \rightarrow BaC_2O_4(s)$

$Ba^{2+}(aq) + SO_4^{2-}(aq) \rightarrow BaSO_4(s)$

GENERAL CONCLUSIONS

Use the information in the following table to answer items **1–5**. The table gives the solubilities of the listed compounds in moles of anhydrous compound that can be dissolved in 1 L of water at 20°C. According to the table, NaNO₃, NaCl, and AgNO₃ are soluble ionic compounds. Their aqueous solutions contain hydrated ions that can be present in high concentrations at room temperature.

Compound	Formula	Solubility at 20°C (mol/L)
silver nitrate	AgNO₃	13
silver chloride	AgCl	1×10^{-5}
sodium nitrate	NaNO₃	10.3
sodium chloride	NaCl	6.2

EXPERIMENT B14 continued

1. **Organizing Information** Suppose that at room temperature you mix fairly concentrated solutions of NaCl and AgNO$_3$. What four ionic species are present in this mixture of two ionic compounds?

   Na$^+$(aq), Cl$^-$(aq), Ag$^+$(aq), NO$_3^-$(aq)

2. **Analyzing Information** When the fairly concentrated solutions of NaCl and AgNO$_3$ are mixed, what compound is likely to precipitate out of solution? Explain why.

   AgCl; the concentrations of Ag$^+$ and Cl$^-$ ions in solution exceed the solubility of AgCl, which is only 1×10^{-5} mol/L at 20°C. Excess Ag$^+$ and Cl$^-$ ions separate from the solution as a precipitate of solid AgCl.

3. **Organizing Conclusions** Write the overall ionic equation for the precipitation reaction described in item **2**.

   Na$^+$(aq) + Cl$^-$(aq) + Ag$^+$(aq) + NO$_3^-$(aq) →

   AgCl(s) + Na$^+$(aq) + NO$_3^-$(aq)

4. **Organizing Conclusions** The overall ionic equation in item **3** shows that the Na$^+$(aq) and NO$_3^-$(aq) ions take no part in the reaction. What is the name given to ions that take no part in a chemical reaction?

   spectator ions

5. **Organizing Conclusions** Write the net ionic equation for the reaction between NaCl and AgNO$_3$.

   Ag$^+$(aq) + Cl$^-$(aq) → AgCl(s)

EXPERIMENT B15

Hydronium Ion Concentration and pH

OBJECTIVES

Recommended time:
50 min

- **Use** pH paper and standard colors to determine the pH of a solution.
- **Determine** hydronium ion concentrations from experimental data.
- **Describe** the effect of dilution on the pH of acids and sodium hydroxide.
- **Relate** pH to the acidity and basicity of solutions.

INTRODUCTION

Solution/Material Preparation

1. Wear safety goggles, a face shield, impermeable gloves, and a lab apron when you prepare the NaOH, HCl, H_3PO_4, NH_3, and CH_3COOH solutions. For all except NaOH and H_3PO_4, work in a hood known to be in operating condition, with another person present nearby to call for help in case of an emergency. Be sure you are within a 30 s walk of a safety shower and eyewash station known to be in operating condition.

2. To prepare 1 L of 0.033 M phosphoric acid, observe the required safety precautions. Slowly and with stirring, add 2 mL of concentrated H_3PO_4 to enough distilled water to make 1 L of solution.

3. To prepare 1 L of 0.1 M acetic acid, observe the required safety precautions. Slowly and with stirring, add 6 mL of glacial acetic acid to enough distilled water to make 1 L of solution.

The concentration of hydronium ions is usually represented by the symbol $[H_3O^+]$, which means the hydronium ion concentration in moles per liter of solution. A one molar (1.0 M) solution of hydronium ions contains 1 mol, 19 g, of hydronium ions per liter of solution. For pure water at 25°C, the hydronium ion concentration, $[H_3O^+]$, is 1.0×10^{-7} M or 10^{-7} M.

In an acidic solution, the hydronium ion concentration is *greater* than 1×10^{-7} M. For example, the H_3O^+ concentration of a 0.00001 M hydrochloric acid (HCl) solution is 1×10^{-5} M.

In all basic solutions, the hydronium ion concentration is *less* than 1×10^{-7} M. The hydronium ion concentration in a basic solution can be determined from the equation for K_w.

$$\text{Because } K_w = [H_3O^+] \times [OH^-] = 1 \times 10^{-14} \text{ mol}^2/L^2$$

$$\text{then } [H_3O^+] = \frac{1 \times 10^{-14} \text{ mol}^2/L^2}{[OH^-]}$$

The concentration of H_3O^+ in a 0.00001 M NaOH basic solution is

$$[H_3O^+] = \frac{1 \times 10^{-14} \text{ mol}^2/L^2}{[OH^-]} = \frac{1 \times 10^{-14} \text{ mol}^2/L^2}{0.00001 \text{ mol/L}}$$

$$= \frac{1 \times 10^{-14} \text{ mol}^2/L^2}{1 \times 10^{-5} \text{ mol/L}} = 1 \times 10^9 \text{ M}$$

The pH of a solution is defined as the common logarithm of the inverse (reciprocal) of the hydronium ion (H_3O^+) concentration. For a 1×10^{-5} M HCl solution,

$$pH = \log \frac{1}{1 \times 10^{-5}} = \log 1 \times 10^5 = 5$$

Because 1×10^{-7} M is the $[H_3O^+]$ in neutral water, a pH of 7 is neutral. Acidic solutions have $[H_3O^+]$ greater than 1×10^{-7} M. The pH values of acids are less than 7. The smaller the pH value, the larger the hydronium ion concentration.

EXPERIMENT B15 continued

4. To prepare 100 mL of 0.1 M hydrochloric acid, observe the required safety precautions. Slowly and with stirring, add 0.9 mL of concentrated HCl to enough distilled water to make 100 mL of solution.

For a 1×10^{-5} M NaOH solution,

$$[H_3O^+] = \frac{1 \times 10^{-14} \text{ mol}^2/\text{L}^2}{1 \times 10^{-5} \text{ mol/L}} = 1 \times 10^{-9} \text{ M}$$

and $\text{pH} = \log \dfrac{1}{1 \times 10^{-9}} = \log 1 \times 10^9 = 9$

Values for pH greater than 7 indicate a basic solution. The higher the pH above the value of 7, the stronger the base and the smaller the hydronium ion concentration.

SAFETY

5. To prepare 125 mL of 0.1 M sodium hydroxide, observe the required safety precautions. Slowly and with stirring, add 0.5 g of NaOH to enough distilled water to make 125 mL of solution.

6. To prepare 250 mL of 0.1 M aqueous ammonia, observe the required safety precautions. Slowly and with stirring, add 1.75 mL of concentrated $NH_3(aq)$ to enough distilled water to make 250 mL of solution.

Always wear safety goggles and a lab apron to protect your eyes and clothing. If you get a chemical in your eyes, immediately flush the chemical out at the eyewash station while calling to your teacher. Know the location of the emergency lab shower and eyewash station and the procedure for using them.

Do not touch any chemicals. If you get a chemical on your skin or clothing, wash the chemical off at the sink while calling to your teacher. Make sure you carefully read the labels and follow the precautions on all containers of chemicals that you use. If there are no precautions stated on the label, ask your teacher what precautions you should follow. Do not taste any chemicals or items used in the laboratory. Never return leftovers to their original containers; take only small amounts to avoid wasting supplies.

Call your teacher in the event of a spill. Spills should be cleaned up promptly, according to your teacher's directions.

MATERIALS

7. To prepare 100 mL of 0.1 M sodium chloride, add 0.6 g of NaCl to enough distilled water to make 100 mL of solution.

- 24-well microplate
- thin-stemmed pipets, 9
- 0.033 M H_3PO_4
- 0.10 M CH_3COOH
- 0.10 M HCl
- 0.10 M NaCl
- 0.10 M Na_2CO_3
- 0.10 M $NaHCO_3$
- 0.10 M NaOH
- 0.10 M $NH_3(aq)$
- 0.10 M NH_4CH_3COO
- distilled water
- pH papers, wide and narrow range

PROCEDURE

8. To prepare 100 mL of 0.1 M sodium carbonate, add 1.1 g of Na_2CO_3 to enough distilled water to make 100 mL of solution.

9. To prepare 100 mL of 0.1 M sodium hydrogen carbonate, add 0.8 g of $NaHCO_3$ to enough distilled water to make 100 mL of solution.

1. Find the pH of each of the nine solutions by applying one drop from the thin-stemmed pipet first to the wide-range pH paper and then to the narrow-range paper. Compare the color produced by each solution to the colors on the chart included with the pH paper. Record your results in Data Table 1.

2. Put 1 drop of 0.10 M HCl into the upper left well of the 24-well plate. Add 9 drops of distilled water. Mix. This new solution is 0.01 M HCl.

3. Transfer 1 drop of the 0.01 M HCl to the next well. Add 9 drops of distilled water. Mix. This solution is 0.001 M HCl. Repeat this process one more time to obtain 0.0001 M HCl.

EXPERIMENT B15 continued

10. To prepare 100 mL of 0.1 M ammonium acetate, add 0.8 g of NH₄CH₃COO to enough distilled water to make 100 mL of solution.

11. Dispense the nine solutions in the thin-stemmed pipets.

Required Precautions

- Safety goggles and a lab apron must be worn at all times.
- Read all safety precautions, and discuss them with your students.
- In case of a spill, use a dampened cloth or paper towel (or more than one towel if necessary) to mop up the spill. Then rinse the towel in running water at the sink, wring it out until it is only damp, and put it in the trash. In case of an acid or base spill, dilute the spill with water first, and then proceed as described.

Techniques to Demonstrate

Show students how to test the solutions and compare the test paper with the pH wide-range and narrow-range standards. Caution them to avoid poor results by thoroughly cleaning and drying the stirring rod between tests.

QUESTIONS

4. Test each concentration of HCl by dipping a clean, dry stirring rod into the solution and touching it first to the wide-range pH paper and then to the narrow-range pH paper. Record your results in Data Table 2.

5. Repeat steps 2 through 4 with 0.10 M NaOH.

6. Repeat steps 2 through 4 with 0.10 M CH₃COOH.

Cleanup and Disposal

7. Clean all apparatus and your lab station. Return equipment to its proper place. Dispose of chemicals and solutions in the containers designated by your teacher. Do not pour any chemicals down the drain or in the trash unless your teacher directs you to do so. Wash your hands thoroughly before you leave the lab and after all work is finished.

Data Table 1

0.1 M solution	pH	0.1 M solution	pH
HCl	1	NaCl	7
CH₃COOH	3	Na₂CO₃	11
H₃PO₄	1.5	NaHCO₃	8
NaOH	13	NH₄CH₃COO	7
NH₃	11		

Data Table 2

Concentration (M)	HCl (pH)	NaOH (pH)	CH₃COOH (pH)
0.10	Students' answers will vary.		
0.010			
0.0010			
0.00010			

1. **Organizing Data** List the solutions in order of decreasing acid strength using your results from step **2**.

HCl, H₃PO₄, CH₃COOH, NH₄CH₃COO, NaCl, NaHCO₃, NH₃, Na₂CO₃, NaOH

EXPERIMENT B15 continued

Pre-lab Discussion

Students may need some discussion of K_w and its relationship to pH.

Disposal

Combine all solutions containing H_3PO_4 and adjust the pH to between 5 and 9 using 1.0 M KOH. (Do not use NaOH because the EPA does not want sodium hydrogen phosphates added to the water.) Then pour down the drain. Combine all other solutions and adjust the pH to between 6 and 8, and pour down the drain.

GENERAL CONCLUSIONS

2. Calculate the pH values for the concentrations prepared in steps **2–4**.

Calculated pH

HCl		NaOH	
0.10 M	1	0.10 M	13
0.010 M	2	0.010 M	12
0.0010 M	3	0.0010 M	11
0.00010 M	4	0.00010 M	10

3. **Analyzing Results** What effect does dilution have on the pH of
   (a) an acid?
   (b) a base?

   a. The pH of an acid is increased as the solution is made more dilute.

   b. The pH of a base is decreased as the solution is made more dilute.

1. **Predicting Outcomes** Solutions of pH 12 or greater dissolve hair. Would a cotton shirt or a wool shirt be affected more by a spill of 0.1 M sodium hydroxide? Explain.

   Wool is animal hair. A shirt made from it will therefore dissolve in sodium hydroxide solution.

2. **Applying Ideas** Read the labels of products found in your home, and identify three basic solutions and three acidic solutions.

   Students' answers will vary but could include the following.

   Basic: soap, some drain cleaners, oven cleaner, ammonia, glass cleaners, bleach, baking soda, washing soda

   Acidic: soda pop, vinegar, citrus juices, aspirin, sour milk, some drain cleaners, muriatic acid

Name _____

Date _____ Class _____

HOLT ChemFile
LAB PROGRAM

EXPERIMENT B16

Titration of an Acid with a Base

OBJECTIVES

Recommended time:
50 min

- **Calibrate** pipets for measuring solutions with precision and accuracy.
- **Recognize** the end point of a titration.
- **Describe** the procedure for standardizing a solution.
- **Determine** the molarity of a base.

INTRODUCTION

Solution/Material Preparation

1. To prepare 200 mL of 0.500 M hydrochloric acid, use a fresh bottle of reagent grade concentrated HCl. Observe the required safety precautions. Assuming that the concentrated HCl is 12 M, slowly and with stirring, add 8.6 mL to enough distilled water to make 200 mL of solution.

2. To prepare 50 mL of 0.6 M sodium hydroxide solution, observe the required safety precautions. Add 1.2 g of NaOH with stirring to enough distilled water to make 50 mL of solution.

3. To prepare 100 mL of phenolphthalein solution, dissolve 1.0 g of phenolphthalein in 50 mL of denatured alcohol and add 50 mL of water.

Titration is a process in which the concentration of a solution is determined by measuring the volume of that solution needed to react completely with a standard solution of known volume and concentration. The process consists of the gradual addition of the standard solution to a measured quantity of the solution of unknown concentration until the number of moles of hydronium ion, H_3O^+, equals the number of moles of hydroxide ion, OH^-. The point at which equal numbers of moles of acid and base are present is known as the equivalence point. An indicator is used to signal when the equivalence point is reached. The chosen indicator must change color very near or at the equivalence point. The point at which an indicator changes color is called the end point of the titration. Phenolphthalein is an appropriate indicator choice for this titration. In acidic solution, phenolphthalein is colorless, and in basic solution, it is pink.

At the equivalence point, the number of moles of acid equals the number of moles of base.

$$\text{mol of } H_3O^+ = \text{mol of } OH^-$$

At the neutral point, the concentrations of H_3O^+ and OH^- are equal. From this relationship you can derive the relationship that is the basis for this experiment, assuming a one-to-one mole ratio.

$$\text{molarity of acid} \times \text{volume of acid} = \text{molarity of base} \times \text{volume of base}$$

In this experiment, you will be given a standard HCl solution and will be told what its concentration is. You will carefully measure a volume of it and determine how much of the NaOH solution of unknown molarity is needed to neutralize the acid sample. From the data you obtain, you can calculate the molarity of the NaOH solution.

SAFETY

Always wear safety goggles and a lab apron to protect your eyes and clothing. If you get a chemical in your eyes, immediately flush the chemical out at the eyewash station while calling to your teacher. Know the location of the emergency lab shower and eyewash station and the procedure for using them.

EXPERIMENT B16 continued

Required Precautions

- Read all safety precautions, and discuss them with your students.
- Safety goggles and a lab apron must be worn at all times.

Do not touch or taste any chemicals. If you get a chemical on your skin or clothing, wash it off at the sink while calling to your teacher. Make sure you carefully read the labels and follow the precautions on all containers of chemicals that you use. If there are no precautions stated on the label, ask your teacher what precautions you should follow. Never return leftovers to their original containers; take only small amounts to avoid wasting supplies.

Call your teacher in the event of a spill. Spills should be cleaned up promptly, according to your teacher's directions.

Never put broken glass into a regular waste container. Broken glass should be disposed of properly in the broken-glass waste container.

MATERIALS

- 10 mL graduated cylinder
- 24-well microplate
- thin-stemmed pipets
- 0.500 M HCl
- phenolphthalein indicator
- NaOH solution of unknown molarity

PROCEDURE

- In case of an acid or base spill, dilute with water first. Then, as with other spills, use a dampened cloth or paper towel (or more than one towel if necessary) to mop up the spill. Rinse the towel in running water at the sink, wring it out until it is only damp, and put it in the trash.
- Wear safety goggles, a face shield, impermeable gloves, and a lab apron when you prepare the NaOH and HCl solutions. For preparing HCl, work in a hood known to be in operating condition, with another person present nearby to call for help in case of an emergency. Be sure you are within a 30 s walk of a safety shower and eyewash station known to be in operating condition.

1. To accurately measure volumes of solution using a pipet, the pipet must be calibrated. Calibration is the comparison of one measuring device with another one. The comparison must be done more than once to reduce error. To calibrate the pipet that will contain the acid, put 5.0 mL of water into the 10 mL graduated cylinder. Fill the thin-stemmed pipet with water from the faucet. Hold the pipet vertical, and add 20 drops of water from the pipet to the water in the cylinder as shown in Figure A. Record the initial volume and the new volume to the nearest 0.01 mL in the Micro Calibration Data Table for trial 1.

FIGURE A

2. Without emptying the cylinder, add 20 more drops from the pipet to the cylinder. Record the volume for this trial. Repeat this step for a third trial. Label the pipet as the acid pipet.

3. Repeat steps **1** and **2** using the pipet that will contain the base. Record your data in the appropriate spaces in the Micro Calibration Data Table.

4. Use your data to find the total volume of all the drops of water added to the cylinder by the acid pipet in the three trials. Record this number in the table.

5. Calculate the average volume of each drop by dividing the total volume by the total number of drops used. Record the average volume in the table.

EXPERIMENT B16 continued

Techniques to Demonstrate

Emphasize the need to hold the pipet consistently in a vertical position when releasing drops. Only then can the calibration be meaningful because when the pipet is in different orientations, the size of the drops is not consistent.

Pre-lab Discussion

Discuss the role of the indicator and the meaning of the terms *end point* and *equivalence point*. Indicators change colors at different pH values. It is important to choose an indicator that changes color at a pH close to the pH of the equivalence point of the titration. For strong bases, such as NaOH, titrated with strong acids, such as HCl, phenolphthalein is a good indicator because the pH of its end point (near pH 7) is very close to the equivalence point.

If students need help with the equations in the Introduction, it may help to work through a set of sample titration data.

Disposal

Combine all waste solutions. Adjust the pH to between 6 and 8 using 1 M HCl or KOH solutions. Then pour the mixture down the drain.

6. Repeat steps **4** and **5** for the base pipet, and record your results.

7. Using the acid pipet, put 20 drops of 0.500 M HCl and 1 drop of phenolphthalein indicator into each of three wells of the 24-well plate. Record the number of drops added. Place the plate on a white sheet of paper in order to more easily detect the color change.

8. Using the base pipet, add NaOH solution of unknown concentration drop by drop to the first well containing HCl. Keep track of the number of drops you are adding. Swirl between each addition. Continue to add drops until the pink color lasts for at least 30 s. Record the number of drops of NaOH added to the first well.

9. Repeat step **8** for the second and third trials, and record your results.

Cleanup and Disposal

10. Clean all apparatus and your lab station. Return equipment to its proper place. Dispose of chemicals and solutions in the containers designated by your teacher. Do not pour any chemicals down the drain or in the trash unless your teacher directs you to do so. Wash your hands thoroughly before you leave the lab and after all work is finished.

Micro Calibration Data Table

Graduated Cylinder Readings (mL)

Trial Number	From acid pipet initial	From acid pipet final	From base pipet initial	From base pipet final
1	5.00	5.64	5.00	5.65
2	5.64	6.30	5.65	6.31
3	6.30	6.92	6.31	6.98

Total volume of drops from acid pipet: __1.92 mL__

Average volume of each drop from acid pipet: __0.032 mL__

Total volume of drops from base pipet: __1.98 mL__

Average volume of each drop from base pipet: __0.033 mL__

CALCULATIONS

1. **Organizing Data** Calculate the volumes of acid used in the three trials. Show your calculations.

    Trial 1: Volume of HCl = __20 drops × 0.032 mL/drop = 0.64 mL__

    Trial 2: Volume of HCl = __20 drops × 0.032 mL/drop = 0.64 mL__

    Trial 3: Volume of HCl = __20 drops × 0.032 mL/drop = 0.64 mL__

EXPERIMENT B16 continued

2. **Organizing Data** Calculate the volumes of base used in the three trials. Show your calculations.

   Trial 1: Volume of NaOH = 16 drops × 0.033 mL/drop = 0.53 mL

   Trial 2: Volume of NaOH = 15 drops × 0.033 mL/drop = 0.50 mL

   Trial 3: Volume of NaOH = 16 drops × 0.033 mL/drop = 0.53 mL

3. **Organizing Data** Determine the moles of acid used in each of the three trials.

   Trial 1: Moles of acid = 0.500 M × 0.00064 L = 0.00032 mol

   Trial 2: Moles of acid = 0.00032 mol

   Trial 3: Moles of acid = 0.00032 mol

4. **Relating Ideas** Write the balanced equation for the reaction between HCl and NaOH.

   HCl + NaOH → NaCl + H$_2$O

5. **Organizing Ideas** Use the mole ratio in the balanced equation and the moles of acid from item **3** to determine the moles of base neutralized in each trial.

   Trial 1: Moles of acid = 0.00032 mol = moles of base

   Trial 2: Moles of acid = 0.00032 mol = moles of base

   Trial 3: Moles of acid = 0.00032 mol = moles of base

6. **Organizing Data** Calculate the molarity of the base for each trial.

   Trial 1: Molarity NaOH = $\dfrac{\text{mol NaOH}}{V \text{ NaOH}} = \dfrac{0.00032 \text{ mol}}{0.00053 \text{ L}} = 0.604$ M

   Trial 2: Molarity NaOH = $\dfrac{0.00032 \text{ mol}}{0.00050 \text{ L}} = 0.640$ M

   Trial 3: Molarity NaOH = $\dfrac{0.00032 \text{ mol}}{0.00053 \text{ L}} = 0.604$ M

7. **Organizing Conclusions** Calculate the average molarity of the base.

   Average molarity of NaOH = $\dfrac{0.604 \text{ M} + 0.640 \text{ M} + 0.604 \text{ M}}{3}$

   = 0.616 M

EXPERIMENT B17

Percentage of Acetic Acid in Vinegar

OBJECTIVES

Recommended time: 50 min

- **Determine** the end point of an acid-base titration.
- **Observe** accurately the quantities of solution in pipets.
- **Calculate** the molarity of vinegar using experimental data.
- **Calculate** the percentage of acetic acid in vinegar.

INTRODUCTION

Materials per lab group
- NaOH solution, standardized, 2.0 mL
- phenolphthalein indicator, 1 mL
- white vinegar, 2.0 mL
- 10 mL graduated cylinder

When sweet apple cider is fermented in the absence of oxygen, the product is an acid, vinegar. Most commercial vinegars are made by fermentation, but some, such as the white vinegar you will use in this experiment, are obtained by the dilution of 100 percent acetic acid. The usual mass percentage of acetic acid in vinegar is between 4.0 percent and 5.5 percent regardless of how it is produced.

The quantity of acetic acid in a sample of vinegar may be found by titrating the sample against a standard basic solution. In this experiment, you will determine the volume of sodium hydroxide solution of known molarity needed to neutralize a measured quantity of a vinegar. From your results, you can calculate the molarity of the vinegar.

SAFETY

- 24-well plate
- thin-stemmed pipets, 2

Solution/Material Preparation

1. Wear safety goggles, face shield, impermeable gloves, and a lab apron if you prepare the NaOH solution or HCl for standardizing the NaOH. Work in a hood known to be in operating condition, with another person present nearby to call for help in case of an emergency. Be sure you are within a 30 s walk of a safety shower and eyewash station known to be in operating condition.

Always wear safety goggles and a lab apron to protect your eyes and clothing. If you get a chemical in your eyes, immediately flush the chemical out at the eyewash station while calling to your teacher. Know the location of the emergency lab shower and eyewash stations and the procedure for using them.

Do not touch any chemicals. If you get a chemical on your skin or clothing, wash the chemical off at the sink while calling to your teacher. Make sure you carefully read the labels and follow the precautions on all containers of chemicals that you use. If there are no precautions stated on the label, ask your teacher what precautions you should follow. Do not taste any chemicals or items used in the laboratory. Never return leftovers to their original containers; take only small amounts to avoid wasting supplies.

Call your teacher in the event of a spill. Spills should be cleaned up promptly, according to your teacher's directions.

Never put broken glass in a regular waste container. Broken glass should be disposed of properly in the broken-glass waste container.

EXPERIMENT B17 continued

MATERIALS

- 10 mL graduated cylinder
- 24-well plate
- thin-stemmed pipets, 2
- NaOH solution, standardized
- NaOH solution, standardized
- white vinegar

PROCEDURE

2. The NaOH solution that students use should be less than 1.0 M. Directions are given for preparing approximately 0.6 M NaOH for standardization. Observe the required safety precautions. Add 1.2 g of NaOH, with stirring, to enough distilled water to make 50 mL of solution. Use reagent grade NaOH.

3. If you or your students are to standardize the NaOH solution, prepare 0.100 M HCl using a fresh bottle of concentrated HCl, preferably one that shows the actual assay of HCl rather than an average assay. Observe the required safety precautions. Assuming that the concentrated HCl is 12 M, slowly and with stirring, add 0.83 mL to enough water to make 100 mL of solution. Adjust the exact volume of concentrated HCl you use depending on its assay.

4. To prepare 100 mL of phenolphthalein indictor, dissolve 1.0 g of phenolphthalein in 50 mL of denatured alcohol, and add 50 mL of water.

5. Avoid cider vinegar because the color will make the recognition of the end point more difficult.

1. To calibrate the pipet you will use for the vinegar, put 5.0 mL of water into the 10 mL graduated cylinder. Read the exact volume of water in the cylinder and record this volume as the initial reading for the acid pipet, trial 1, in the Calibration Data Table. Fill the thin-stemmed pipet with water from the faucet. Hold the pipet vertically, and transfer 20 drops of water from the pipet to the graduated cylinder. Read and record the new volume as the final volume for trial 1. Without emptying the cylinder, add 20 more drops and record the volume for this second trial. Repeat for a third trial. Label the pipet *Acid*.

2. Repeat step **1** for the second pipet. Record the data, and label this pipet *NaOH*.

Calibration Data Table

Readings in Graduated Cylinder (mL)

Trial	From acid pipet initial	From acid pipet final	From base pipet initial	From base pipet final
1	5.00	5.65	5.00	5.64
2	5.65	6.31	5.64	6.30
3	6.31	6.98	6.30	6.92

Total volume of drops from acid pipet: __1.98 mL__

Average volume of each drop of acid: __0.033 mL__

Total volume of drops from base pipet: __1.92 mL__

Average volume of each drop of base: __0.032 mL__

3. Hold the pipets vertically each time you use them. Add 20 drops of vinegar and one drop of phenolphthalein to each of three wells of the 24-well plate.

4. To the first well, add standardized sodium hydroxide solution drop by drop from a vertical pipet, swirling the plate gently after each drop. Continue until the pink color persists for 30 s. Record the number of drops added in the Data Table.

5. Repeat step **4** with the two other 20-drop samples of vinegar. Perform each titration in the same way, titrating to the same shade of pink each time. Refill the sodium hydroxide pipet, if necessary. Record your results.

Cleanup and Disposal

6. Clean all apparatus and your lab station. Return equipment to its proper place. Dispose of chemicals and solutions in the containers designated by your teacher. Do not pour any chemicals down the drain or in the trash unless your teacher directs you to do so. Wash your hands thoroughly before you leave the lab and after all work is finished.

EXPERIMENT B17 continued

Required Precautions

- Safety goggles and a lab apron must be worn at all times.
- Read all safety precautions, and discuss them with your students.

Data Table
Number of drops added to well

Trial	Vinegar	NaOH
1	20	19
2	20	20
3	20	19

CALCULATIONS

- Students should not handle concentrated acid or base solutions.
- Use commercially available distilled white vinegar instead of diluting glacial acetic acid.
- In case of an acid or base spill, dilute first with water. Then while wearing disposable plastic gloves, mop up the spill with wet cloths or a wet cloth mop designated for spill cleanup.

Techniques to Demonstrate

Show students how to hold the pipet in a vertical position every time they use the pipets.

Pre-lab Discussion

Review the stoichiometry of acid-base reactions and relate it to the end point of the titration and the definition of molarity. Many students may need help with the calculations. It might help to work through a set of sample data.

Disposal

Combine all waste solutions. Adjust the pH to between 6 and 8 using 1 M HCl or KOH solutions. Then pour the mixture down the drain.

1. **Organizing Data** Calculate the volumes of vinegar and NaOH used for each of the three trials.

    Volume of vinegar = **20 drops × 0.033 mL/drop = 0.66 mL**

    Volume of NaOH:
    Trial 1 = **19 drops × 0.032 mL/drop = 0.61 mL**

    Trial 2 = **20 drops × 0.032 mL/drop = 0.64 mL**

    Trial 3 = **19 drops × 0.032 mL/drop = 0.61 mL**

2. **Organizing Data** Using the molarity of the standardized NaOH solution you used, determine the moles of NaOH used in each of the three trials.

    Molarity NaOH = **0.623**

    Trial 1 = **0.623 mol/L × 0.00061 L = 0.00038 mol NaOH**

    Trial 2 = **0.623 mol/L × 0.00064 L = 0.00040 mol NaOH**

    Trial 3 = **0.623 mol/L × 0.00061 L = 0.00038 mol NaOH**

3. **Organizing Ideas** Write the balanced equation for the reaction between vinegar and sodium hydroxide. (Hint: the formula for acetic acid is CH_3COOH.)

    $CH_3COOH + NaOH \rightarrow H_2O + NaCH_3COO$

4. **Organizing Data** Use the results of your calculations in item **2** and the mole ratio in the equation in item **3** to determine the moles of base used to neutralize the vinegar (acid) in each trial.

    Trial 1 = **0.00038 mol NaOH × $\dfrac{1 \text{ mol vinegar}}{1 \text{ mol NaOH}}$ = 0.00038 mol vinegar**

    Trial 2 = **0.00040 mol NaOH × $\dfrac{1 \text{ mol vinegar}}{1 \text{ mol NaOH}}$ = 0.00040 mol vinegar**

    Trial 3 = **0.00038 mol NaOH × $\dfrac{1 \text{ mol vinegar}}{1 \text{ mol NaOH}}$ = 0.00038 mol vinegar**

ChemFile

EXPERIMENT B17 continued

5. **Organizing Data** Use the moles of acid calculated in item **4** and the volumes of the acid used for each trial to calculate the molarities of vinegar for the three trials.

$$\text{Trial 1} = \frac{0.00038 \text{ mol}}{0.00066 \text{ L}} = 0.58 \text{ M}$$

$$\text{Trial 2} = \frac{0.00040 \text{ mol}}{0.00066 \text{ L}} = 0.61 \text{ M}$$

$$\text{Trial 3} = \frac{0.00038 \text{ mol}}{0.00066 \text{ L}} = 0.58 \text{ M}$$

6. **Organizing Data** Calculate the average molarity of the vinegar.

$$\frac{0.58 \text{ M} + 0.61 \text{ M} + 0.58 \text{ M}}{3} = 0.59 \text{ M}$$

7. **Organizing Ideas** Use the periodic table to calculate the mass of one mole of acetic acid, CH_3COOH.

$$\begin{aligned} H &= 1.01 \text{ g} \times 4 = 4.04 \text{ g} \\ C &= 12.01 \text{ g} \times 2 = 24.02 \text{ g} \\ O &= 16.00 \text{ g} \times 2 = 32.00 \text{ g} \\ & \phantom{= 16.00 \text{ g} \times 2 =\ } 60.06 \text{ g} \end{aligned}$$

8. **Organizing Data** Use the average molarity for your vinegar sample to determine the mass of CH_3COOH in 1 L of vinegar.

$$\frac{\text{g } CH_3COOH}{1 \text{ L vinegar}} = \frac{0.59 \text{ mol } CH_3COOH}{1.0 \text{ L vinegar}} \times \frac{60.06 \text{ g } CH_3COOH}{1 \text{ mol } CH_3COOH}$$

$$= \frac{35 \text{ g } CH_3COOH}{1 \text{ L vinegar}}$$

9. **Organizing Conclusions** Assume that the density of vinegar is very close to 1.00 g/mL so that the mass of 1 L of vinegar is 1000 g. Calculate the percentage of acetic acid in your vinegar sample. (Hint: the mass of acetic acid in 1 L of vinegar, calculated in item **8**, divided by the total mass of vinegar in a liter and multiplied by 100 gives the percentage of acetic acid in vinegar.)

$$\% \ CH_3COOH = \frac{35 \text{ g } CH_3COOH}{1000 \text{ g vinegar}} \times 100 = 3.5\%$$

EXPERIMENT B17 continued

GENERAL CONCLUSIONS

1. **Applying Conclusions** Why is it important for a company manufacturing vinegar to regularly check the molarity of its product?

   Consumers expect a certain degree of sourness (acidity) in the vinegar they buy. They would not be satisfied if the acidity changed from one bottle to the next.

2. **Analyzing Methods** What was the purpose of the phenolphthalein? Could you have titrated the vinegar sample without the phenolphthalein?

   The change in the color of phenolphthalein showed when the titration was complete. Without that signal, it would not have been possible to know when enough NaOH had been added to neutralize the acetic acid.

3. **Analyzing Methods** Why was it important to hold the pipets vertically each time you used them?

   Consistently holding the pipets in a vertical position helped ensure that the drops were the same size.

4. **Evaluating Data** Share your data with other lab groups, and calculate a class average for the molarity of the vinegar. Compare the class average with your results, and calculate the average deviation and the standard deviation.

   Students' answers will vary.

ChemFile

EXPERIMENT B17 **103**

EXPERIMENT B17 continued

5. **Designing Experiments** What possible sources of error can you identify in this procedure? If you can think of ways to eliminate them, ask your teacher to approve your plan, and run the procedure again.

   Students' answers will vary. Be sure that any procedures are safe.

6. **Relating Ideas** Explain the difference between the equivalence point and the end point. How are they different? Can they be the same?

   The end point is the point at which the indicator used in a titration changes color. The equivalence point is the point at which the number of moles of acid (H^+) equals the number of moles of base (OH^-). If the indicator is chosen properly, its color change will occur near the equivalence point. They could be the same.

Name _____

Date _____ Class _____

ChemFile
LAB PROGRAM

EXPERIMENT B18

Clock Reactions

OBJECTIVES

Recommended time:
50 min

- **Observe** chemical processes and interactions.
- **Compare** chemical reactions by measuring reaction rates.
- **Relate** reaction rate concepts to observations.
- **Infer** a conclusion from experimental data.

INTRODUCTION

Many medications are advertised as having a timed-release action. When the medication is ingested, it becomes active in the body over time. Pharmaceutical companies can control how quickly a drug enters the bloodstream controlling the rate at which the drug is digested in the stomach. In this experiment you will look at how temperature, surface area, and pH affect the rate at which an effervescent antacid tablet reacts with a solvent.

SAFETY

Solution/Material Preparation

1. Wear goggles, face shield, impermeable gloves, and a lab apron when you prepare the HCl. Work in a functioning hood within a 30 s walk to a functioning safety shower and eyewash station.

2. To prepare 0.10 M HCl, add 8.3 mL of concentrated HCl to enough distilled water to make 1.00 L of solution. Add the acid slowly, while stirring to avoid overheating.

Always wear safety goggles and a lab apron to protect your eyes and clothing. If you get a chemical in your eyes, immediately flush the chemical out at the eyewash station while calling to your teacher. Know the location of the emergency lab shower and eyewash station and the procedure for using them.

Do not touch any chemicals. If you get a chemical on your skin or clothing, wash the chemical off at the sink while calling to your teacher. Make sure you carefully read the labels and follow the precautions on all containers of chemicals that you use. If there are no precautions stated on the label, ask your teacher what precautions you should follow. Do not taste any chemicals used in the laboratory. Never return leftovers to their original containers; take only small amounts to avoid wasting supplies.

Call your teacher in the event of a spill. Spills should be cleaned up promptly, according to your teacher's directions.

MATERIALS

3. You may wish to divide the class into groups, with each group using a different brand of effervescent tablet.

- effervescent tablets, 24
- 0.10 M HCl, 60 mL
- ice (optional)
- balance
- 50 mL beaker
- beaker tongs
- 25 mL graduated cylinder

ChemFile   EXPERIMENT B18   **105**

HRW material copyrighted under notice appearing earlier in this work.

EXPERIMENT B18 continued

4. The test tubes should be 18 mm × 150 mm or larger.

5. You may want to use a hot plate to heat water and an ice bath for cold water, if the water from the tap does not vary from room temperature.

PROCEDURE

Required Precautions

- Safety goggles and a lab apron must be worn at all times.
- Read all safety precautions, and discuss them with your students.
- In case of a spill, use a dampened cloth or paper towel to mop up the spill. Then rinse the towel in running water at the sink, wring it out until it is damp, and put it in the trash.

Techniques to Demonstrate

You can skip the calibration step if you demonstrate the dissolution described in step 1.

Pre-lab Discussion

Discuss the importance of using a consistent method for determining the start and stop times for the reaction. Determination of the reaction time is subjective and may vary a great deal among the lab groups.

Disposal

All solutions may be flushed down the drain.

- mortar and pestle
- spatula
- stirring rod
- timing device or stopwatch
- test tubes, 4
- test tube holder and rack
- thermometer, nonmercury

1. This is a calibration step to determine how long the reaction takes before you start your actual testing. Begin by measuring 15 mL of water, using the graduated cylinder. Add one effervescent tablet to the beaker of water. Start timing the reaction when you add the tablet. When you can no longer see any visible reaction, note the time elapsed.

2. Fill 4 test tubes each with 15 mL of cold water from the tap. Record the temperature of the water in the Data Table.

3. Get three effervescent tablets. Break two of the tablets in half. Break one of the halves in half again.

4. Add 1/4 tablet to the first test tube. Start timing, and record the reaction rate in the Data Table.

5. Repeat step **4** using 1/2 tablet in test tube 2, 3/4 tablet in test tube 3, and 1 whole tablet in test tube 4. Record the reaction rates for each test tube in the Data Table. Pour the solutions down the drain, and clean the test tubes for the next set of tests.

6. Repeat steps **2–5** with water at room temperature (about 25°C).

7. Repeat steps **2–5** with hot water from the tap.

8. Use room temperature water for this set of reactions. Repeat steps **2–5** but crush each of the samples using a mortar and pestle before adding them to the test tubes.

9. Repeat steps **2–5**, but crush each sample, use room temperature water in the test tubes, and stir the mixture.

10. Repeat steps **2–5**, but crush each sample, use cold water in the test tubes, and stir the mixture.

11. Repeat steps **2–5**, but crush each sample, use hot water in the test tubes, and stir the mixture.

12. Repeat steps **2–5**, but crush each sample, use 0.1 M HCl in the test tubes, and stir the mixture.

Cleanup and Disposal

13. Clean all apparatus and your lab station. Return equipment to its proper place. Dispose of chemicals and solutions in the containers designated by your teacher. Do not pour any chemicals down the drain or in the trash unless your teacher directs you to do so. Wash your hands thoroughly before you leave the lab and after all your work is finished.

EXPERIMENT B18 continued

Data Table

	Solvent temperature (°C)	Time 1/4 tablet	Time 1/2 tablet	Time 3/4 tablet	Time whole tablet
Cold water	Temperatures will vary.	52 s	57 s	66 s	78 s
Room-temperature water		45 s	49 s	55 s	63 s
Hot water		34 s	36 s	39 s	46 s
Crushed tablet, room-temperature water		30 s	33 s	38 s	44 s
Crushed tablet, room-temperature water, stirred		22 s	24 s	27 s	31 s
Crushed tablet, cold water, stirred		47 s	52 s	60 s	73 s
Crushed tablet, hot water, stirred		15 s	17 s	21 s	26 s
Crushed tablet, 0.1 M HCl, stirred		12 s	15 s	17 s	22 s

QUESTIONS

1. **Analyzing Results** Which mixture had the fastest reaction time? Which mixture had the slowest reaction time?

   The crushed 1/4 tablet in 0.1 M HCl had the fastest reaction time. The whole tablet reacting with cold water had the slowest reaction time.

2. **Analyzing Methods** What was the effect of increased temperature on the reaction rate?

   Increasing the temperature caused an increase in all reaction rates. The cold-water reactions were the slowest.

3. **Evaluating Conclusions** Which of the following reaction-rate variables is being tested in this experiment: temperature, catalyst, concentration, surface area, or nature of reactants?

   Temperature, surface area, and nature of the reactants are all being tested.

EXPERIMENT B18 continued

4. **Inferring Conclusions** Effervescent tablets generally contain a carbonate compound that decomposes to produce CO_2. Ask your teacher for the product label to determine the active ingredient in your sample. Write an equation for the reaction that produces CO_2.

Answers will vary depending on the product used.

GENERAL CONCLUSIONS

1. **Evaluating Methods** What are some likely sources of error in this experiment? If you can think of ways to eliminate them, ask your teacher to approve your suggestion, and run more trials.

Students' answers will vary but may include the following:

difficulty in making each fractional sample the same size and

imprecision of time measurements. Students may suggest

minimizing these errors by repeated trials or by using a

standardized mass for each of the fractional samples. Be sure

any new procedures are carefully planned and safe.

2. **Predicting Outcomes and Designing Experiments** Would all brands of effervescent tablets show the same reaction rates? Design an experiment to test your answer. If your teacher approves your suggestion, perform the experiment and record your results.

Students' answers will vary, depending upon the products used.

3. **Formulating Conclusions** You are working for a pharmaceutical company. Your job is to write the labeling for a bottle of indigestion-relief tablets. Suggest directions for the label to help customers obtain the fastest relief.

Answers will vary; one possible option is "For best results, break

up the tablets, and place them in a glass of warm water. Stir the

mixture with a spoon until the tablets are dissolved."

Name _____

Date _____ Class _____

EXPERIMENT B19

Equilibrium

OBJECTIVES

Recommended time: 50 min; each part can be performed separately

- **Observe** color changes in solutions as indications of shifts in equilibrium.
- **Explain** shifts in equilibrium by applying Le Châtelier's principle.
- **Observe** and **explain** the common-ion effect.

INTRODUCTION

Materials (for quantities not in student's list)

- 0.1 M $CuSO_4$, 3 mL
- 0.025 M CH_3COOH, 25 mL
- 0.025 M $FeCl_3$, 1 mL
- 0.025 M KSCN, 1 mL
- 1.0 M HCl, 1 mL
- 1.0 M NH_3, 2 mL
- $Fe(NO_3)_3$, 0.5 g
- K_2HPO_4, 0.5 g
- KCl, 0.5 g
- methyl red indicator solution, 1 mL

Some chemical reactions run to completion and are commonly referred to as *end reactions*. These are reactions in which a product is essentially undissociated, given off as a gas, or precipitated. Many other reactions do not run to completion. The products of these reactions do not leave the field of action but react with each other to reform the original reactants. An equilibrium is established in which the forward and reverse actions continue at equal rates so that there is no net change in the quantities of reactants or products.

A reaction at equilibrium is affected by both concentration and temperature. Le Châtelier's principle states that a system in equilibrium tends to shift to relieve any stress placed on it by a change in concentration or temperature. For example, the addition of more reactants will shift an equilibrium in the direction of producing more products, thus relieving the stress on the reactants. In this experiment, you will investigate some systems at equilibrium and interpret your observations in terms of Le Châtelier's principle.

SAFETY

- CH_3COONa, 1 g
- NH_4SCN, 0.5 g

Always wear safety goggles and a lab apron to protect your eyes and clothing. If you get a chemical in your eyes, immediately flush the chemical out at the eyewash station while calling to your teacher. Know the location of the emergency lab shower and eyewash and the procedure for using them.

Solution/Material Preparation

1. Wear safety goggles, a face shield, impermeable gloves, and a lab apron when you prepare the HCl, NH_3, and iron(III) chloride solution containing HCl. Work in a hood known to be in operating condition, with another person present nearby to call for help in case of an emergency. Be sure you are within a 30 s walk of a safety shower and eyewash station known to be in operating condition.

Do not touch any chemicals. If you get a chemical on your skin or clothing, wash the chemical off at the sink while calling your teacher. Make sure you carefully read the labels and follow the precautions on all containers of chemicals that you use. If there are no precautions stated on the label, ask your teacher what precautions you should follow. Do not taste any chemicals or items used in the laboratory. Never return leftovers to their original containers; take only small amounts to avoid wasting supplies.

Call your teacher in the event of a spill. Spills should be cleaned up promptly, according to your teacher's directions.

ChemFile

EXPERIMENT B19 **109**

EXPERIMENT B19 continued

Never put broken glass in a regular waste container. Broken glass should be disposed of properly in the broken-glass waste container.

MATERIALS

2. To prepare 100 mL of copper(II) sulfate, dissolve 2.5 g of $CuSO_4 \cdot 5H_2O$ in enough water to make 100 mL of solution.

3. For reasons of safety, do not dilute glacial acetic acid to prepare 0.025 M acetic acid. Instead, dilute white vinegar 33-fold (30 mL of vinegar diluted to 1 L).

- 24-well microplate
- glass stirring rod
- test tube, 13 mm × 100 mm
- thin-stemmed pipets, 8
- 0.1 M $CuSO_4$
- 0.025 M CH_3COOH
- 0.025 M $FeCl_3$
- 0.025 M KSCN
- 1.0 M HCl
- 1.0 M NH_3
- $Fe(NO_3)_3$
- K_2HPO_4
- KCl
- methyl red indicator solution
- CH_3COONa
- NH_4SCN

PROCEDURE

4. To prepare 100 mL of 0.025 M iron(III) chloride, observe the required safety precautions. Slowly and while stirring, add 2 mL of concentrated HCl to about 80 mL of water. Add 0.67 g of $FeCl_3 \cdot 6H_2O$ and then dilute to 100 mL.

5. To prepare 1 L of 0.025 M potassium thiocyanate, add 2.4 g of KSCN to enough water to make 1 L of solution.

6. To prepare 50 mL of 1.0 M hydrochloric acid, observe the required safety precautions. Slowly and while stirring, add 4.3 mL of concentrated HCl to enough water to make 50 mL of solution.

7. To prepare 100 mL of 1.0 M aqueous ammonia solution, observe the required safety precautions. Slowly and while stirring, add 7.6 mL of concentrated ammonia water to enough water to make 100 mL of solution.

1. Put the 24-well plate on a sheet of white paper so the color changes will be more easily seen, as shown in Figure A.

FIGURE A

(Top row wells labeled: CH_3COOH + methyl red + CH_3COONa)
(Fifth row wells labeled: $FeCl_3$ / KSCN)

Part 1 Common-ion effect

2. Put 10 drops of 0.025 M acetic acid into a well in the top row of the plate. Add one drop of methyl red indicator solution. Mix the solution.

   Observations: A violet-red color appears.

EXPERIMENT B19 continued

8. To prepare 100 mL of methyl red indicator, add 0.5 g of dry methyl red to 100 mL of methanol.

Required Precautions

- Safety goggles and a lab apron must be worn at all times.
- Read all safety precautions, and discuss them with your students.
- In case of a spill, use a dampened cloth or paper towel (or more than one towel if necessary) to mop up the spill. Then rinse the towel in running water at the sink, wring it out until it is only damp, and put it in the trash. In the event of an acid or base spill, or acidic $FeCl_3$, dilute the spill first with water and then proceed as described above.

Pre-lab Discussion

To help students interpret their observations, encourage them to refer to the equilibrium equation for each reaction as they apply the various stresses. Students should be asking themselves what reactant or product is being affected by the stress and what affect the change in that reactant or product has on the equilibrium.

Disposal

Combine all solutions. Add sufficient 1.0 M NaOH to precipitate all of the copper as copper(II) hydroxide. Adjust the pH to between 8 and 10 and pour the mixture down the drain.

3. Add one small crystal of CH_3COONa and mix to dissolve.

Observations: An orange to yellow color appears.

Part 2 Le Châtelier's principle

4. Put 3.0 mL of 0.1 M $CuSO_4$ in a test tube. The ion in the solution is $Cu(H_2O)_4^{2+}$.

Observations: The solution is light blue.

5. Add drops of 1.0 M NH_3 until the color changes and intensifies. The ion in the solution is $Cu(NH_3)_6^{2+}$.

Observations: The solution is dark blue.

6. Add drops of 1.0 M HCl until the color changes again.

Observations: The solution returns to very light blue, the same color as the original solution.

7. Explain the solution's appearance after each addition in terms of Le Châtelier's principle.

As ammonia is added, the water molecules in the $Cu(H_2O)_4^{2+}$ ion are replaced by NH_3 molecules.

$Cu(H_2O)_4^{2+} + 6NH_3 \rightleftarrows Cu(NH_3)_6^{2+} + 4H_2O$

The addition of HCl favors the reverse reaction because HCl reacts with the ammonia, which is a reactant in the equation.

Part 3 Complex-ion equilibrium

8. Mix 5 drops each of 0.025 M $FeCl_3$ and 0.025 M KSCN in a single well in the fifth row of the microplate. Mix thoroughly.

Observations: A deep-red color appears.

ChemFile EXPERIMENT B19 **111**

EXPERIMENT B19 continued

9. Put 1 drop of the solution made in step **8** into each of the four wells of the sixth row. Dilute each with 20 drops of distilled water. Add the following to the wells. Well *1*: a crystal of $Fe(NO_3)_3$; Well *2*: a crystal of NH_4SCN; Well *3*: a crystal of KCl; Well *4*: a crystal of K_2HPO_4.

Observations:

1. color darkens (deep red)

2. color darkens (deep red)

3. no change

4. solution turns colorless

Cleanup and Disposal

10. Clean all apparatus and your lab station. Return equipment to its proper place. Dispose of chemicals and solutions in the containers designated by your teacher. Do not pour any chemicals down the drain or in the trash unless your teacher directs you to do so. Wash your hands thoroughly before you leave the lab and after all work is finished.

QUESTIONS

1. **Relating Ideas** Methyl red indicator has a pH range of 4.2 (red-violet color) to 6.2 (yellow color). Explain the color change that occurred in Part 1 of the procedure, when solid CH_3COONa was added to the acetic acid solution.

   When sodium acetate dissolved in the solution, the acetate ions reacted with the H^+ ions and lowered the H^+ ion concentration.

2. **Analyzing Information**
   a. What color is $Cu(H_2O)_4^{2+}$? light blue
   b. What color is $Cu(NH_3)_6^{2+}$? deep blue

3. **Analyzing Information**
   a. When ammonia was added to the light-blue copper complex in step **5**, what ion was formed?
   b. Write the balanced equation for this reaction.
   c. Use Le Châtelier's principle to explain why the addition of HCl caused the deep-blue solution to change to light blue.

   a. $Cu(NH_3)_6^{2+}$

   b. $Cu(H_2O)_4^{2+} + 6NH_3 \rightleftarrows Cu(NH_3)_6^{2+} + 4H_2O$

   c. HCl reacts with the NH_3 on the reactant side of the equation, removing it as NH_4^+ ions. This stress causes the equilibrium to shift to the left to produce more ammonia and more of the light-blue complex ion.

EXPERIMENT B19 continued

4. **Relating Ideas** The Fe^{3+} ion and the SCN^- ion form the complex $FeSCN^{2+}$ ion, which is a deep-red color. Write the balanced equation for the formation of the $FeSCN^{2+}$ ion.

$Fe^{3+} + SCN^- \leftrightarrow FeSCN^{2+}$

5. **Inferring Conclusions** Indicate whether the product increased, decreased, or remained unchanged when the equilibrium conditions were changed in the following ways:
   a. Fe^{3+} ions were added (well *1*).
   b. SCN^- ions were added (well *2*).
   c. KCl was added (well *3*).

   a. Equilibrium shifts to the right creating more $FeSCN^{2+}$, the product.

   b. Equilibrium shifts to the right creating more $FeSCN^{2+}$, the product.

   c. no effect

GENERAL CONCLUSIONS

1. **Inferring Conclusions** Some sunglasses darken when exposed to bright sunlight and become more transparent when they encounter shade. These sunglasses are made from a glass that contains small crystals of silver chloride. When photons of ultraviolet light hit these transparent crystals, the silver chloride changes into dark silver and chlorine atoms.

$$AgCl \xrightleftharpoons{light} Ag^0 + Cl^0$$

In the shade, the equilibrium reverses, producing transparent silver chloride crystals again. This equilibrium, however, does not occur in solutions. When a freshly prepared precipitate of silver chloride is exposed to sunlight, the surface turns black. Although this is the same reaction as in the photosensitive sunglasses, this time it is not reversible. Explain why. (Hint: How are the environments of the products different in the two situations?)

In the sunglasses, the silver and chlorine atoms are trapped in the structure of the glass. Consequently, they can recombine. In the precipitate, the chlorine atoms combine to form chlorine gas and leave the system.

EXPERIMENT B19 continued

2. **Analyzing Conclusions** Many changes took place in the three parts of this experiment. Write a single statement to summarize all of these changes.

When there is a change in the concentration of any product or reactant in a system that is in equilibrium, the equilibrium will be shifted in whichever direction will counteract the change.

EXPERIMENT B20

Oxidation-Reduction Reactions

OBJECTIVES

Recommended time: 50 min

- **Observe** oxidation-reduction reactions between metals.
- **Describe** typical oxidation-reduction reactions.
- **Determine** relative strengths of some oxidizing and reducing agents.

INTRODUCTION

Solution/Material Preparation

1. To prepare 100 mL of 0.1 M copper(II) nitrate, dissolve 2.4 g of $Cu(NO_3)_2 \cdot 3H_2O$ in enough distilled water to make 100 mL of solution.

2. To prepare 100 mL of 0.1 M zinc nitrate, dissolve 3.0 g of $Zn(NO_3)_2$ in enough distilled water to make 100 mL of solution.

A substance that loses electrons during a chemical reaction is said to be oxidized. A substance that gains electrons is said to be reduced. If one reactant gains electrons, another must lose an equal number. Thus, oxidation and reduction must occur simultaneously and to a comparable degree.

The stronger the tendency of a species to take electrons, the greater is its strength as an oxidizing agent and the more easily it is reduced. The stronger the tendency of a species to give up electrons, the greater is its strength as a reducing agent and the more readily it is oxidized. The silver ion, Ag^+, has a strong tendency to acquire an electron to form the silver atom, Ag. Thus, the Ag^+ ion is a strong oxidizing agent. In this experiment, you will determine the relative strengths of some metals as reducing agents and the relative strengths of their ions as oxidizing agents.

SAFETY

3. To prepare 100 mL of 0.1 M magnesium nitrate, dissolve 1.8 g of $Mg(NO_3)_2 \cdot 2H_2O$ in enough distilled water to make 100 mL of solution.

4. To prepare 25 mL of 0.1 M iron(III) chloride, observe the required safety precautions. Slowly and while stirring, add 1.0 mL of concentrated HCl to about 4 mL of distilled water. Then add 0.68 g of $FeCl_3 \cdot 6H_2O$ and dilute the solution to a final volume of 25 mL.

Always wear safety goggles and a lab apron to protect for your eyes and clothing. If you get a chemical in your eyes, immediately flush the chemical out at the eyewash station while calling to your teacher. Know the location of the emergency lab shower and eyewash and the procedures for using them.

Do not touch any chemicals. If you get a chemical on your skin or clothing, wash the chemical off at the sink while calling to your teacher. Make sure you carefully read the labels and follow the precautions on all containers of chemicals that you use. If there are no precautions stated on the label, ask your teacher what precautions you should follow. Do not taste any chemicals or items used in the laboratory. Never return leftovers to their original containers; take only small amounts to avoid wasting supplies.

Call your teacher in the event of a spill. Spills should be cleaned up promptly, according to your teacher's directions.

ChemFile   EXPERIMENT B20   **115**

EXPERIMENT B20 continued

MATERIALS

5. To prepare 100 mL of 0.1 M potassium permanganate, add 1.6 g of $KMnO_4$ to enough distilled water to make 100 mL of solution.

- 24-well microplate
- thin-stemmed pipets, 6
- 0.1 M $Cu(NO_3)_2$
- 0.1 M $FeCl_3$
- 0.1 M $KMnO_4$
- 0.1 M $Mg(NO_3)_2$
- 0.1 M $Zn(NO_3)_2$
- 1.0 M H_2SO_4
- 0.5 cm copper wire, 3 pieces
- iron(II) sulfate
- 5 cm magnesium ribbon
- 0.5 cm zinc strip, 3 pieces
- tin(II) chloride

PROCEDURE

6. To prepare 50 mL of 1.0 M sulfuric acid, observe the required safety precautions. Slowly and with stirring, add 2.8 mL of concentrated H_2SO_4 to enough distilled water to make 50 mL of solution.

Required Precautions

- Read all safety precautions, and discuss them with your students.

- Safety goggles and a lab apron must be worn at all times.

- In case of a spill, use a dampened cloth or paper towel (or more than one towel if necessary) to mop up the spill. Then rinse the towel in running water at the sink, wring it out until it is only damp, and put it in the trash. In case of an acid or base spill, dilute the spill first with water and then follow the procedure as described.

- Wear safety goggles, a face shield, impermeable gloves, and a lab apron when you prepare the 1.0 M H_2SO_4 and the iron(III) chloride solution containing HCl. Work in a hood known to be in operating condition, with another person present nearby to call for help in case of an emergency. Be sure you are within a 30 s walk of a safety shower and eyewash station known to be in operating condition.

1. Place 5 drops of copper(II) nitrate solution in each of the four wells in the top row of a microplate, as shown in the Figure A. Place 5 drops of zinc nitrate in each of the four wells in the next row. Place 5 drops of magnesium nitrate in each of the four wells in the third row.

FIGURE A

2. Add a small piece of copper wire to each of the solutions in the first column of the microplate. Add a small piece of zinc to the solutions in the second column of the plate and a small piece of magnesium to the solutions in the third column. The fourth column will be used for comparison.

3. Put the microplate on a white sheet of paper and observe for several minutes. Record the results in your Data Table.

4. Put 5 drops of iron(III) chloride solution in one of the empty wells. Add one small crystal of tin(II) chloride to this solution. Record the results in the Data Table.

5. Add a small crystal of iron(II) sulfate to another of the empty wells. Add 10 drops of water and 5 drops of 1.0 M sulfuric acid. Add 0.1 M potassium permanganate solution dropwise, mixing after each addition. Continue adding until two color changes have occurred. Record the results in the Data Table.

EXPERIMENT B20 continued

Pre-lab Discussion

- Encourage students to think in terms of changing oxidation states. Make sure they understand that when a metal dissolves in water, it has become an ion and therefore has lost electrons (been oxidized). Conversely, if a metallic ion gains electrons (is reduced), it is no longer soluble in water and will precipitate. Emphasize that strong oxidizing agents are readily reduced and strong reducing agents are readily oxidized.

Cleanup and Disposal

6. Clean all apparatus and your lab station. Return equipment to its proper place. Dispose of chemicals and solutions in the containers designated by your teacher. Do not pour any chemicals down the drain or in the trash unless your teacher directs you to do so. Wash your hands thoroughly before you leave the lab and after all work is finished.

Data Table

	Cu	Zn	Mg
$Cu(NO_3)_2$	NR	Zn dissolves, Cu ppt	Mg dissolves, Cu ppt
$Zn(NO_3)_2$	NR	NR	Mg dissolves, Zn ppt
$Mg(NO_3)_2$	NR	NR	NR
$FeCl_3 + SnCl_2$	color changes from orange to light green		
$KMnO_4 + FeSO_4$	green to orange to deep purple color		

QUESTIONS

Disposal

Save the metal strips for reuse unless they are too corroded, in which case, you should put them in the trash. Combine all solutions. If the mixture is purple or even slightly pink in color add 1.0 M $Na_2S_2O_3$ until decolorized. Then adjust the pH to be between 5 and 9, dilute tenfold, and pour down the drain.

1. **Organizing Data** Which metal was oxidized by two other ions?

   magnesium

2. **Organizing Data** Which metal was oxidized by only one other ion?

   zinc

3. **Organizing Data** Which metal was not oxidized by any of the ions?

   copper

4. **Analyzing Results** Arrange the three metals in order of their relative strengths as reducing agents, placing the strongest first.

   1. Mg^0, Zn^0, Cu^0

5. **Analyzing Results** Arrange the three metallic ions in order of their relative strengths as oxidizing agents, placing the strongest first. Write the reduction half-reaction for each ion.

   1. $Cu^{2+} + 2e^- \rightarrow Cu^0$

   2. $Zn^{2+} + 2e^- \rightarrow Zn^0$

   3. $Mg^{2+} + 2e^- \rightarrow Mg^0$

ChemFile

EXPERIMENT B20

EXPERIMENT B20 continued

6. **Inferring Results** Copper is oxidized in the presence of silver ions. The net ionic reaction is the following.

$$Cu + 2Ag^+ \rightarrow Cu^{2+} + 2Ag$$

Write net ionic equations for the following:
a. the reaction of copper and zinc

$Cu^{2+} + Zn \rightarrow Cu^0 + Zn^{2+}$

b. the reaction of zinc and magnesium

$Zn^{2+} + Mg^0 \rightarrow Zn^0 + Mg^{2+}$

c. the reaction of copper and magnesium. (Hint: Use your answers to Questions 4 and 5 to determine which metal is oxidized and which ion is reduced.)

$Cu^{2+} + Mg^0 \rightarrow Cu^0 + Mg^{2+}$

7. **Analyzing Results** In step 4, the Fe^{3+} ion was reduced to the Fe^{2+} ion.
a. What was the reducing agent?

Sn^{2+} ion

b. What change did the Sn^{2+} ion undergo?

Sn^{2+} was oxidized to Sn^{4+}

c. Write the net ionic equation for the overall reaction:
$2Fe^{3+} + 6Cl^- + Sn^{2+} + 2Cl^- \rightarrow 2Fe^{2+} + 4Cl^- + Sn^{4+} + 4Cl^-$

$2Fe^{3+} + Sn^{2+} \rightarrow 2Fe^{2+} + Sn^{4+}$

8. **Analyzing Results** The permanganate ion, MnO_4^-, which is purple in color, is a strong oxidizing agent. The manganese(II) ion, Mn^{2+}, is practically colorless. What occurred during the addition of potassium permanganate to the Fe^{2+} ions?

The almost colorless (pale green) solution of Fe^{2+} changed to the yellow-brown of Fe^{3+}.

GENERAL CONCLUSIONS

1. **Predicting Outcomes** What would happen to the metals if iron nails were used to secure sheets of copper to a roof?

Iron is a more active metal than copper. Thus, the nails will corrode, causing the sheets of copper to become loose.

EXPERIMENT B21

Cathodic Protection: Factors that Affect the Corrosion of Iron

OBJECTIVES

Recommended time: 50 min with about 20 min the day before and 10 min the day after the lab

- **Identify** factors that affect the rate of corrosion of iron.
- **Use** a control for comparison.
- **Recognize** corrosion of iron as the result of oxidation-reduction.
- **Describe** ways of reducing corrosion.

INTRODUCTION

Solutions/Material Preparation

1. To prepare 100 mL of 0.1 M potassium hexacyanoferrate(III), dissolve 3.4 g of $K_3Fe(CN)_6$ in enough water to make 100 mL of solution.

2. To prepare 100 mL of phenolphthalein indicator, dissolve 1.0 g of phenolphthalein in 50 mL of denatured alcohol, and add 50 mL of water.

3. To prepare 125 mL of 0.1 M sodium hydroxide, observe the required safety precautions. Dissolve 0.5 g of NaOH in enough water to make 125 mL of solution.

4. To prepare 100 mL of 0.1 M sodium chloride, dissolve 0.53 g of NaCl in enough water to make 100 mL of solution.

Corrosion is a chemical reaction in which a metal is oxidized. Iron corrodes in the presence of oxygen and water. The metal is converted to brittle metal oxides called rust. Bicycles, car bodies, tools, and appliances become useless and must be replaced when they corrode.

The exact nature of the corrosion process is not well understood, but it is known to be an oxidation-reduction reaction in which iron is initially converted to iron(II) ions.

$$Fe(s) \rightarrow Fe^{2+}(aq) + 2e^-$$

At the same time, hydroxide ions (OH^-) are formed from water and oxygen molecules.

$$1/2 O_2(g) + H_2O(l) + 2e^- \rightarrow 2OH^-(aq)$$

Adding the two equations gives the following oxidation-reduction equation.

$$Fe(s) + 1/2 O_2(g) + H_2O(l) \rightarrow Fe^{2+}(aq) + 2OH^-(aq)$$

Further reaction with oxygen and water produces rust, so the presence of Fe^{2+} ions and OH^- ions is evidence that corrosion has occurred. The formation of a blue precipitate when $K_3Fe(CN)_6$ is added to a solution indicates the presence of Fe^{2+} ions. Phenolphthalein indicator solution turns pink in the presence of OH^- ions. In this experiment, you will use $K_3Fe(CN)_6$ and phenolphthalein to verify the presence of Fe^{2+} and OH^- ions as you investigate some of the factors that affect the process of corrosion.

SAFETY

Always wear safety goggles and a lab apron to provide protection for your eyes and clothing. If you get a chemical in your eyes, immediately flush the chemical out at the eyewash station while calling to your teacher. Know the location of the emergency lab shower and eyewash station and the procedure for using them.

EXPERIMENT B21 continued

5. To prepare 100 mL of 0.1 M sodium carbonate, dissolve 1.1 g of Na_2CO_3 in enough water to make 100 mL of solution.

6. To prepare 50 mL of 0.1 M sodium triphosphate, dissolve 1.9 g of $Na_3PO_4 \cdot 12H_2O$ in enough water to make 50 mL of solution.

7. To prepare 50 mL of 0.1 M potassium nitrate, dissolve 0.5 g of KNO_3 in enough water to make 50 mL of solution.

8. To prepare 50 mL of 0.1 M sodium oxalate, dissolve 0.65 g of $Na_2C_2O_4$ in enough water to make 50 mL of solution.

Do not touch any chemicals. If you get a chemical on your skin or clothing, wash the chemical off at the sink while calling to your teacher. Make sure you carefully read the labels and follow the precautions on all containers of chemicals that you use. If there are no precautions stated on the label, ask your teacher what precautions you should follow. Do not taste any chemicals or items used in the laboratory. Never return leftovers to their original containers; take only small amounts to avoid wasting supplies.

Call your teacher in the event of a spill. Spills should be cleaned up promptly, according to your teacher's directions.

When using a Bunsen burner, confine long hair and loose clothing. If your clothing catches on fire, WALK to the emergency lab shower, and use it to put out the fire. **Do not heat glassware that is broken, chipped, or cracked.** Use tongs or a hot mitt to handle heated glassware and other equipment because hot glassware does not always look hot.

Never put broken glass into a regular waste container. Broken glass should be disposed of properly in the broken-glass waste container.

MATERIALS

9. To prepare 100 mL of 0.1 M hydrochloric acid, observe the required safety precautions. Slowly and while stirring, add 0.86 mL of concentrated HCl to enough water to make 100 mL of solution.

10. To prepare 1 L of 0.1 M nitric acid, observe the required safety precautions. Slowly and while stirring, add 6.4 mL of concentrated HNO_3 to enough water to make 1 L of solution.

11. To prepare 200 mL of 0.1 M sulfuric acid, observe the required safety precautions. Slowly and while stirring, add 1.1 mL of concentrated H_2SO_4 to enough water to make 200 mL of solution.

Required Precautions

- Read all safety precautions, and discuss them with your students.

- balance
- 24-well microplate
- 400 mL beaker
- Bunsen burner and heating set-up
- glass stirring rod
- petri dishes, 2
- pliers
- thin-stemmed pipets, 11
- tongs
- wire gauze, ceramic center
- 0.1 M $K_3Fe(CN)_6$
- powdered agar-agar
- copper strip or wire
- iron nails, 4
- iron wire
- magnesium wire
- phenolphthalein indicator
- wide-range pH paper
- test solutions, 10
  0.1 M HCl
  0.1 M HNO_3
  0.1 M KNO_3
  0.1 M Na_2CO_3
  0.1 M NaCl
  0.1 M NaOH
  0.1 M $Na_2C_2O_4$
  0.1 M Na_3PO_4
  0.1 M H_2SO_4
  distilled water

EXPERIMENT B21 continued

PROCEDURE

- Safety goggles and a lab apron must be worn at all times.

- In case of a spill, use a dampened cloth or paper towel (or more than one towel if necessary) to mop up the spill. Then rinse the towel in running water at the sink, wring it out until it is only damp, and put it in the trash. In the event of an acid or base spill, dilute the spill with water first and then proceed as described.

- Wear safety goggles, a face shield, impermeable gloves, and a lab apron when you prepare the NaOH, HCl, HNO₃, and H₂SO₄ solutions. While preparing the acids, work in a hood known to be in operating condition, with another person present nearby to call for help in case of an emergency. Be sure you are within a 30 s walk of a safety shower and eyewash station known to be in operating condition.

Techniques to Demonstrate

Show students how to use the pliers to bend the nails. Caution them to use beaker tongs or hot mitts when moving the beaker containing the suspension of agar-agar.

Pre-lab Discussion

Remind students of the Prussian blue test for iron ions, and review how the phenolphthalein indicator distinguishes between acids and bases.

Disposal

The metals may be saved for reuse. If they are too corroded, put them in the trash. Combine all solutions and adjust the pH to be between 5 and 9, then dilute 100-fold and pour down the drain.

Part 1

1. Determine the pH of the test solutions by placing a drop of each on a strip of wide-range pH paper. Record your results in the Data Table.

2. Put 10 drops of each test solution into a separate well of the microplate. Label the microplate wells or make a diagram showing the position of each solution.

3. Add a 0.5 cm piece of iron wire to each well. Cover the plate with plastic wrap, and set it aside until the next day. Wash your hands.

4. On the following day, place the plate on a sheet of white paper, and observe any changes that have taken place. Record your results in the Data Table.

5. Add 1 drop of 0.1 M $K_3Fe(CN)_6$ to each of the wells. Record the results.

6. Dispose of the metal and plastic wrap as your teacher directs you. Clean and rinse the plate. Return the plate and pipets to the location specified by your teacher. Wash your hands.

Part 2

1. In a 400 mL beaker, heat 200 mL of distilled water to boiling. While stirring, add 2 g of agar-agar. Heat and continue to stir until the agar-agar is thoroughly dispersed. Turn off the heat.

2. With tongs, remove the agar-agar mixture from the ring stand. Add 10 drops of 0.1 M $K_3Fe(CN)_6$ and 5 drops of phenolphthalein indicator. Set the beaker aside to cool.

3. While the agar-agar suspension is cooling, prepare four nails. Use the pliers to bend one nail so that it makes a 90° angle. Wrap a second nail with copper wire and a third with magnesium ribbon.

4. Place the straight nail and the bent nail in one of the petri dishes. Place the nail wrapped with copper wire and the nail wrapped with magnesium ribbon in the other petri dish, as shown in Figure A. Be sure the nails do not touch each other. Pour the lukewarm agar-agar over the nails until they are covered.

FIGURE A

5. Cover the dishes, and set them aside until tomorrow. Wash your hands.

EXPERIMENT B21 continued

6. The next day, record your observations in the Data Table. Remember that all parts of the experiment were allowed to react for the same length of time. Therefore, a greater amount of corrosion indicates an increased rate of reaction.

Cleanup and Disposal

7. Clean all apparatus and your lab station. Return equipment to its proper place. Dispose of chemicals and solutions in the containers designated by your teacher. Do not pour any chemicals down the drain or in the trash unless your teacher directs you to do so. Wash your hands thoroughly before you leave the lab and after all work is finished.

Data Table

Solution	pH range	Appearance	$K_3Fe(CN)_6$ test
NaOH	base		
NaCl	no change	corrosion	+
HCl	acid	corrosion	+
Na_2CO_3	base		
H_2SO_4	acid	corrosion	+
$Na_2C_2O_4$	base		
KNO_3	no change	corrosion	+
HNO_3	acid	corrosion	+
Na_3PO_4	base		
Distilled water	no change	corrosion	+

Day 2 Observations:
Students' answers will vary.

QUESTIONS

1. **Organizing Data** Divide the solutions in Part 1 into three categories according to their pH.

Acidic	Neutral	Basic
HCl	NaCl	NaOH
H_2SO_4	KNO_3	$Na_2C_2O_4$
HNO_3	H_2O	Na_2CO_3
		Na_3PO_4

EXPERIMENT B21 continued

2. Analyzing Results In which group of solutions was there the most evidence of corrosion?

the acids

3. Analyzing Results In which group of solutions was there the least evidence of corrosion?

the bases

4. Relating Ideas Use the principles of equilibrium to explain your answers to Questions 2 and 3. (Hint: Refer to the equation in the Introduction. Assuming equilibrium, how would the presence of either H^+ ions or OH^- ions affect the amount of product formed?)

If the corrosion is occurring in a basic environment in which there

are extra OH^- ions, the equilibrium will be shifted to the left.

Fewer $Fe^{2+}(aq)$ and $2OH^-(aq)$ ions will be formed.

5. Analyzing Methods Why was the straight nail included in Part 2 of the experiment?

to serve as a control for comparison with the bent nail

6. Analyzing Results and Relating Ideas Why do you think the head and the point of the nail corroded more than the middle? (Hint: Think about how nails are made.)

The head and the point have been subjected to more stress in

manufacturing.

7. Analyzing Results What regions of the bent nail corroded?

the head, the tip, and the bend

8. Analyzing Ideas Explain why corrosion occurs more easily in these regions.

The nail has been subjected to stress in these regions.

EXPERIMENT B21 continued

9. **Relating Ideas** Find an Activity Series table and use it to explain why the reactions were different when the nail was wrapped in copper and magnesium.

   Magnesium loses electrons more easily than iron, so the

   magnesium corrodes instead of the iron. Copper loses electrons

   less easily than iron, so the iron corrodes.

10. **Relating Ideas** Magnesium or zinc blocks are often attached to the hulls of ships. Explain how this helps to prevent corrosion of the ship.

    Both zinc and magnesium are more active metals than iron and

    will corrode first, or be sacrificed.

GENERAL CONCLUSIONS

1. **Organizing Conclusions** Summarize the factors that promote corrosion.

   acid environment, stress, presence of a less reactive metal

2. **Predicting Outcomes** The installer of a new battery in your car carelessly splashes some of the battery acid on the inside of your car fender. If the fender is made of iron, should you worry about corrosion? Explain.

   yes, because an acid environment promotes corrosion

3. **Inferring Conclusions** List some actions that might be taken to prevent or reduce corrosion of iron.

   - Coat the metal with paint or oil to prevent oxygen and
     moisture from reaching it.
   - Keep the environment basic.
   - Avoid denting and bending the metal.
   - Attach the metal to a more active metal.

Name _____

Date _____ Class _____

HOLT ChemFile
LAB PROGRAM

EXPERIMENT **B22**

Determination of Vitamin C in Citrus Juices

OBJECTIVES

Recommended time: 30 min

- **Determine** the amounts of iodine needed to oxidize the Vitamin C in samples of fruit juice.
- **Compare** the amounts of iodine used to the amount needed to oxidize a standard solution of vitamin C.
- **Calculate** the concentrations of vitamin C in the samples of fruit juice.

INTRODUCTION

Materials (for quantities not in student's list)

- 1% starch solution, 1 mL
- iodine solution, 1.5 mL
- fruit juices, 3, 1 mL each
- vitamin C solution, 1 mg/mL, 1 mL

Solution/Material Preparation

1. To prepare 100 mL of 1% starch solution, dissolve 1 dropperful of commercial liquid laundry starch in 100 mL of water.

Vitamin C is an important nutrient in your diet. It is essential for preventing the disease called scurvy and for maintaining good health. Vitamin C is required daily because it is water-soluble and cannot be stored by the body. Fortunately, vitamin C is abundant in foods, especially in citrus fruits. In this experiment, you will determine the concentration of vitamin C in three fruit juices by comparing them with a solution of vitamin C having a known concentration. To do so, you will take advantage of the ability of vitamin C to act as a reducing agent.

In cells, vitamin C is involved in a variety of oxidation-reduction reactions. When it reduces other molecules, such as I_2, it becomes oxidized. You will add iodine solution to a solution of vitamin C until a starch indicator turns blue. The oxidation reaction is complex but can be summarized in this way.

$$\text{vitamin C} + I_2 \rightarrow 2I^- + \text{oxidized vitamin C}$$

When all of the vitamin C has been oxidized, the iodine is available to react with starch to form a blue color.

$$I_2 + \text{starch} \rightarrow \text{blue color}$$

In this experiment, you will determine the amount of I_2 needed to oxidize a sample of standard vitamin C solution and then compare it with the amount of iodine needed to oxidize a sample of citrus juice having the same volume. From your data you will calculate the concentration of vitamin C in the fruit juice in mg/mL.

SAFETY

2. To prepare 100 mL of iodine solution, dissolve 0.6 g of KI in 50 mL of water. Dissolve 0.6 g of I_2 in 5 mL of ethanol. Mix the two solutions and dilute to a volume of 100 mL.

Always wear safety goggles and a lab apron to protect your eyes and clothing. If you get a chemical in your eyes, immediately flush the chemical out at the eyewash station while calling to your teacher. Know the location of the emergency lab shower and eyewash station and the procedure for using them.

Do not touch any chemicals. If you get a chemical on your skin or clothing, wash the chemical off at the sink while calling to your teacher. Make sure you carefully read the labels and follow the precautions on all

ChemFile

EXPERIMENT B22 **125**

EXPERIMENT B22 continued

3. A variety of citrus juices may be used. One thin-stemmed pipet will hold enough juice for two or three lab groups. Fresh, canned, frozen, or powdered juices may be used.

containers of chemicals that you use. If there are no precautions stated on the label, ask your teacher what precautions you should follow. Do not taste any chemicals or items used in the laboratory. Never return leftovers to their original containers; take only small amounts to avoid wasting supplies.

Call your teacher in the event of a spill. Spills should be cleaned up promptly, according to your teacher's directions.

MATERIALS

- 24-well microplate
- thin-stemmed pipets, 4
- 1% starch solution
- iodine solution
- fruit juices, 3 different types
- vitamin C solution, 1 mg/mL

PROCEDURE

4. To prepare a standard solution of vitamin C, dissolve a vitamin C tablet in enough water to provide a solution concentration of 1 mg/mL. (Refer to the label on the tablet container for the vitamin C content.) To remove insoluble additives, dissolve the tablet in 50 mL less water than is required. Filter to remove solid material and dilute to the appropriate volume. Do not heat to hasten dissolving because vitamin C decomposes with heat. Prepare the solution as close to the time of use as is practical.

Required Precautions

- Safety goggles and a lab apron must be worn at all times.
- Iodine will stain clothes and hands; therefore, gloves are recommended.
- In case of a spill, use a dampened cloth or paper towel (or more than one towel if necessary) to mop up the spill. Then rinse the towel in running water at the sink, wring it out until it is only damp, and put it in the trash.

1. Put 5 drops of the standard vitamin C solution into each of three wells in the first row of the microplate, as shown in Figure A.

FIGURE A

2. Put 5 drops of fruit juice 1 into each of the three wells in the second row of the plate.

3. Put 5 drops of fruit juice 2 into each of the three wells in the third row of the plate. Repeat this procedure for juice 3, placing it in the wells in the fourth row.

4. Add 1 drop of starch solution to each well.

EXPERIMENT B22 continued

Disposal

Combine all liquids with stirring, and add sufficient 1.0 M $Na_2S_2O_3$ solution to decolorize the starch-iodine complex. Then pour down the drain.

5. Count the drops as you add the iodine solution dropwise to the first of the vitamin C wells in the first row. Continue adding iodine until the solution in the well stays purple-blue for at least 15 s. Record the number of drops in the Data Table.

6. Repeat step 5 for the two other samples of standard vitamin C in row 1, and then do the same for each of the solutions in the wells in rows 2, 3, and 4. Record all your results in the Data Table.

Cleanup and Disposal

7. Clean all apparatus and your lab station. Return equipment to its proper place. Dispose of chemicals and solutions in the containers designated by your teacher. Do not pour any chemicals down the drain or in the trash unless your teacher directs you to do so. Wash your hands thoroughly before you leave the lab and after all work is finished.

Data Table

Solution	Drops of Iodine Solution		
	Trial 1	Trial 2	Trial 3
Vitamin C	18	17	18
Juice 1	12	12	12
Juice 2	9	7	8
Juice 3	11	13	11

CALCULATIONS

1. **Organizing Data** Determine the average number of drops of iodine solution required to oxidize the Vitamin C in the standard solution and in the three juice samples. Record these averages in the calculations table.

   Vitamin C: $\dfrac{18 + 17 + 18}{3}$ = 18 drops

   Juice 1: $\dfrac{12 + 12 + 12}{3}$ = 12 drops

   Juice 2: $\dfrac{9 + 7 + 8}{3}$ = 8 drops

   Juice 3: $\dfrac{11 + 13 + 11}{3}$ = 12 drops

2. **Organizing Data** In item **1,** you calculated the average number of drops needed to oxidize the vitamin C in the standard vitamin C sample and in the three juices. Recall that the concentration of the known solution of vitamin C is 1 mg/mL. You determined the number of drops needed to oxidize a 5 drop-sample with concentration equal to 1 mg/mL. If you assume that all the drops delivered by the four pipets were the same size, then the ratio of the concentration of fruit juice 1 to the concentration of the standard vitamin C solution is equal to the ratio of the drops of iodine solution used to titrate fruit juice 1

EXPERIMENT B22 continued

to the drops needed to titrate the standard vitamin C solution. Use this relationship to calculate the concentrations of vitamin C in the three fruit juices.

$$\frac{\text{concentration juice 1 (mg vit C/mL)}}{1 \text{ mg standard vit C soln/mL}} = \frac{12 \text{ drops}}{18 \text{ drops}}$$

Concentration Juice 1 = 0.67 mg/mL

$$\frac{\text{concentration juice 2 (mg vit C/mL)}}{1 \text{ mg standard vit C soln/mL}} = \frac{8 \text{ drops}}{18 \text{ drops}}$$

Concentration Juice 2 = 0.44 mg/mL

$$\frac{\text{concentration juice 3 (mg vit C/mL)}}{1 \text{ mg standard vit C soln/mL}} = \frac{12 \text{ drops}}{18 \text{ drops}}$$

Concentration Juice 3 = 0.67 mg/mL

GENERAL CONCLUSIONS

1. **Inferring Conclusions** Which of the juices has the highest concentration of Vitamin C/mL of juice?

   Students' answers will vary depending on the juices used.

   For the sample data, juices 1 and 3 are the highest, at

   0.67 mg/mL.

2. **Applying Conclusions** If the minimum daily requirement for vitamin C is 0.060 g/day, what is the minimum amount of juice you must drink each day? (Assume that the juice is your only source of vitamin C.)

   $$\frac{0.060 \text{ g}}{1 \text{ day}} \times \frac{10^3 \text{ mg}}{1 \text{ g}} \times \frac{1 \text{ mL}}{0.67 \text{ mg}} = 88 \text{ mL/day}$$